Electrostatic Phenomena on Planetary Surfaces

Electrostatic Phenomena on Planetary Surfaces

Carlos I Calle PhD
Senior Research Scientist
NASA Kennedy Space Center, USA

Morgan & Claypool Publishers

Rights & Permissions
To obtain permission to re-use copyrighted material from Morgan & Claypool Publishers, please contact info@morganclaypool.com.

ISBN   978-1-6817-4477-3 (ebook)
ISBN   978-1-6817-4476-6 (print)
ISBN   978-1-6817-4479-7 (mobi)

DOI   10.1088/978-1-6817-4477-3

Version: 20170201

IOP Concise Physics
ISSN 2053-2571 (online)
ISSN 2054-7307 (print)

A Morgan & Claypool publication as part of IOP Concise Physics
Published by Morgan & Claypool Publishers, 40 Oak Drive, San Rafael, CA, 94903 USA

IOP Publishing, Temple Circus, Temple Way, Bristol BS1 6HG, UK

To my grandson Liam

Contents

Preface

Our knowledge of planetary environments has increased considerably as a result of the planetary exploration missions that NASA primarily along with ESA and other space agencies has launched in recent decades. Electrostatic phenomena occurring on the solar system bodies are among the most important ones to consider, as they affect the weather in planets and moons with atmospheres and the physical properties of the surfaces on planets and moons lacking an atmosphere. These phenomena are not completely understood even on Earth. Data on electrostatic phenomena from recent missions are currently being analyzed and results are being published and discussed in scientific journals and at conferences.

My aim with this concise book is to provide an overall understanding of the different aspects of electrostatic phenomena as they occur on planetary atmospheres and surfaces, to show the reader what is known in this important field, what is expected from planned exploration missions, and what the big unknowns are. I have included numerous references at the end of each chapter to guide further research into this field.

Acknowledgements

The idea for this concise book on electrostatic phenomena on planetary surfaces came from Nicki Dennis at Morgan & Claypool and Institute of Physics joint publishing. I am indebted to her. I would also like to thank Dr José Nuñez at NASA Kennedy Space Center for his support of this work.

I would like to acknowledge the valuable contributions to our work on electrostatic phenomena on the Moon and Mars by the research team at the NASA Kennedy Space Center Electrostatics and Surface Physics Laboratory. Their work has led to the development of several electrostatics-based technologies for NASA's planetary exploration missions. I would like to acknowledge in particular Dr Charles Buhler, Dr Sid Clements, Paul Mackey, Dr Michael Hogue, Dr James Mantovani, Michael Johansen, Ellen Arens, James Phillips III, and Rachel Cox. Finally, I would like to thank Dr Luz Marina Calle, my wife and fellow NASA scientist. She has always been a wonderful sounding board during my research and writing endeavors.

Author biography

Carlos I Calle

Carlos I Calle is a senior research scientist at NASA Kennedy Space Center and is founder and head of the Electrostatics and Surface Physics Laboratory. He is currently working on the problem of electrostatic phenomena of granular and bulk material as they apply to planetary surfaces, particularly that of Mars, developing instrumentation for future planetary exploration missions. He has over 150 scientific publications, four books, and several patents.

Dr Calle received the NASA Exceptional Technological Achievement Medal in 2010, the NASA Spaceflight Awareness Award in 2003 for his outstanding contributions to the space program, and the NASA Silver Snoopy Award in 2007 for his exceptional contributions to human spaceflight.

Other books by Carlos I Calle:

Superstrings and Other Things: A Guide to Physics 2nd edn (London: CRC Press/Taylor & Francis) 660 pp (2009)

  French edition: *Supercordes et Autres Ficelles: A Voyage au Coeur de la Physique* (Paris : Dunod) 608 pp (2004)

The Universe, Order Without Design (Amherst, NY: Prometheus Books) (2009)

Coffee With Einstein (London: Duncan Baird Publishers) 150 pp (2007)

  German edition: *Auf einen Kaffee mit Einstein* (Munich: Deutscher Taschenbuch Verlag) (2009)

  French edition: *Un café avec Einstein* (Paris: Gründ) (2008)

  Turkish edition: *Hayali Söyleşiler: Einstein* (Istanbul: Kolektif Kitap) (2012)

  Chinese edition: 在咖啡馆遇见爱因斯坦 (Beijing: Wang Zheng Heilongjiang University Press) (2013)

Einstein for Dummies (Hoboken, NJ: Wiley) 374 pp (2005)

  Dutch edition: *Einstein voor Dummies* (Amsterdam: Addison-Wesley) (2005)

  Spanish edition: *Einstein para Dummies* (Bogota: Grupo Editorial Norma) (2006)

  Serbian edition: *Ainštajn za neupućene* (Banja Luka: Mikro knjiga) (2006)

Electrostatic Phenomena on Planetary Surfaces

Carlos I Calle

Chapter 1

Introduction

The diverse planetary environments in the Solar System react in somewhat different ways to the encompassing influence of the Sun. These different interactions define the electrostatic phenomena that take place on and near planetary surfaces. The desire to understand the electrostatic environments of planetary surfaces goes beyond scientific inquiry. These environments have enormous implications for both human and robotic exploration of the Solar System.

In Solar System bodies with an atmosphere, electrostatic phenomena are determined by the physical and chemical properties of those atmospheres. Atmospheric pressure and density, as well as composition, govern the way objects acquire, hold on to, and release electrostatic charge. The transfer of electric charge from an object to another can happen in different ways. If the transfer is fast, as in electrical breakdown, lightning can occur. In addition to Earth, lightning has been detected on Jupiter and Saturn. Recent atmospheric models predict that lightning should also occur on Venus, Mars, and on Saturn's moon Titan. Despite several efforts to detect lightning on these bodies with instrumentation aboard spacecraft, the question remains open. Electrostatic discharges have been detected from Uranus and Neptune, although it is not known if those discharges are caused by lightning.

When the charge transfer is slow, a glow discharge known as a corona can take place. These corona discharges occur on Earth and may also occur on Venus, Mars, Titan, and the three giant planets.

The most extensively studied electrostatic environment is clearly that of the Earth. The Earth's magnetosphere, produced by the interaction of the solar wind with the planet's magnetic field, controls the electrostatic charge content of the atmosphere. The magnetosphere slows down, deflects, and traps many solar wind particles. These trapped particles form the doughnut-shaped regions called Van Allen radiation belts, a nearly impenetrable barrier that prevents the most energetic electrons from reaching the Earth's atmosphere.

Spacecraft and satellites interact with the space plasma environment around the Earth. This interaction generates electrostatic charging on these orbiting craft, a

doi:10.1088/978-1-6817-4477-3ch1   1-1

complex phenomenon that may interfere with their operation and may disrupt or damage power, navigation, communications, and other instrumentation.

In addition to the Earth, Mercury, Saturn, Jupiter, Neptune, and Uranus, as well as Jupiter's moon Ganymede have magnetic fields. As happens on Earth, these magnetic fields interact with the solar wind, producing magnetospheres which act as semipermeable barriers to the solar wind particles. The magnetic field lines plunge at the planets' magnetic poles, allowing particles from the solar wind to reach lower altitudes. Mass ejections from the Sun's corona, commonly known as solar flares, increase the flux of solar wind particles for several hours. These particles interact with the magnetic fields of the planets, releasing trapped particles which trigger reactions with atmospheric molecules that release photons. In the polar regions, these photons form the auroras, known as the northern and southern lights on Earth. Auroras have been observed on Jupiter, Saturn, and Neptune.

The electrostatic environment of airless bodies, such as the Moon, Mercury, and the major asteroids, is the result of the direct interaction of the surface with the solar wind, cosmic rays, and solar radiation. The surfaces of these Solar System bodies develop a charge that balances the sum of all these current fluxes. The electrostatic interaction between the charged surfaces of these Solar System bodies and the surrounding plasma results in the arrangement of the plasma particles in the form of a shield that surrounds the surface. This shield limits the magnitude of the charge that develops on these surfaces.

The surfaces of Mars, Mercury, the Moon, and a few other Solar System bodies, are covered with a layer of fine dust. Interaction of the surface dust with the unmanned exploration rovers on Mars or with rovers and astronaut boots during the Apollo missions to the Moon generates electrostatic charge on the bodies as they repeatedly make contact and separate, a phenomenon called triboelectric charging. Triboelectric charging is unlikely to be a concern on the day side of airless bodies, but can reach levels that may cause electrostatic discharges on the dark sides due to a stronger electron flux in those regions. Triboelectric charging is also a concern on Mars and NASA took steps to mitigate the issue with all its rovers operating on the planet.

Data from recent and current planetary missions from NASA, the European Space Agency, and the Japanese space agency JAXA has provided unprecedented information on the electrostatic environments of Solar System planets and moons. Some of these data are still being analyzed and new discoveries are frequently being made. But many unknowns remain. Not all planned experiments in planetary missions have been successful. Some missions carrying valuable experiments have failed altogether and at least one important experiment on a successful mission was not carried out.

In the pages that follow I describe in some detail what is known about the electrostatic environment of the Solar System from early and current experiments on Earth, as well as what is being learned from the instrumentation on the space exploration missions of the last few decades. But before embarking in this study, I present a brief review of the basic principles of electrostatics.

Electrostatic Phenomena on Planetary Surfaces

Carlos I Calle

Chapter 2

Electrostatics principles

2.1 Coulomb's law and the principle of superposition

A brief introduction to the fundamental principles of electrostatics should be of use before looking at the electrostatic phenomena on the different bodies of the Solar System.

Electrostatics is a component of electromagnetism, a branch of physics that describes the behavior of the electromagnetic force, one of the four fundamental forces in nature. The fundamental problem of electromagnetism is to understand the force that electric charges exert on one another. Electrostatics restricts itself to answer this question when charges are not in motion. As we shall see, electrostatics does allow for short, slow motions of charges as they interact with each other. Continuous motion of charges generates magnetic fields and that takes us beyond electrostatics to the range of electromagnetism.

The solution to the fundamental problem of electromagnetism was given by Charles Coulomb in the eighteenth century and is known today as Coulomb's law. If two charges q_1 and q_2, considered to be *point* charges (that is, charges on an ideal object with no physical extension), are separated by a distance r, the force that q_2 exerts on q_1 is

$$\mathbf{F} = k\frac{q_1 q_2}{r^2}\hat{\mathbf{r}} \tag{2.1}$$

where $\hat{\mathbf{r}}$ is a unit vector along the line joining the two charges and pointing away from q_2 toward q_1. This expression is Coulomb's law. The SI unit of charge is the *coulomb*, C, and the force is given in newtons, N. In SI units, the constant of proportionality k is usually given in terms of the permittivity of free space ε_0, as $1/4\pi\varepsilon_0$, with

$$\varepsilon_0 = 8.85 \times 10^{-12} \ \text{C}^2/(\text{N} \cdot \text{m}^2).$$

doi:10.1088/978-1-6817-4477-3ch2 2-1

The force is repulsive if the charges have equal signs and attractive if they have opposite signs.

The force between these two point charges q_1 and q_2 is not affected by the presence of any other point charges in the vicinity. The total force on q_1 due to all the charges present is calculated by computing the force on q_1 due to q_2 alone, then calculating the force on q_1 due to q_3 alone, and so on, and adding all these contributions. This is known as the *principle of superposition*. Coulomb's law and the principle of superposition are the only two physics concepts required for electrostatics. The rest is mathematical manipulation of these concepts [1].

2.2 The electric field

How is the force between two charges separated by a distance r transmitted from one charge to the other? The classical solution, introduced by Michael Faraday in the nineteenth century, involves the concept of *field*. According to Faraday, the space around an electric charge is distorted in such a way that any other charge placed in this space accelerates with a force that is given by Coulomb's law.

The electric field around a charge q can be mapped by placing a positive charge q_0 at different locations around q. The force at each one of these locations is given by Coulomb's law and points away from q (figure 2.1). This force is

$$\mathbf{F} = \frac{q_0}{4\pi\varepsilon_0}\frac{q}{r^2}\hat{\mathbf{r}}$$

or

$$\mathbf{F} = q_0\mathbf{E}$$

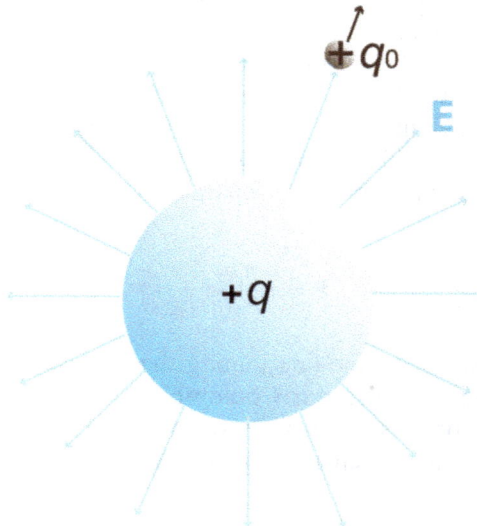

Figure 2.1. The electric field in the space around a positive charge q is mapped by placing a positive charge q_0 at different points around the charge q.

where **E** is the electric field of the charge q:

$$\mathbf{E} = \frac{1}{4\pi\epsilon_0}\frac{q}{r^2}\hat{\mathbf{r}}. \tag{2.2}$$

2.3 Gauss's law

The electric *field lines* around a charge q are the graphical representation of the electric field vectors **E** at different locations around the charge. From (2.2), we can see that the field lines around a positive charge point radially outward from the charge and those around a negative charge point radially toward the charge. If we assume that the number of field lines is proportional to the magnitude of the charge q, we can see that if we place this charge inside a closed surface, all the field lines coming out (for a positive charge) or going in (for a negative charge) cross this closed surface. We can define the *flux* of **E**, ϕ, through any surface as the net number of field lines crossing this surface. Since the number of field lines is proportional to the magnitude of the charge, the flux is

$$\phi = \oint \mathbf{E} \cdot \mathbf{da} = \frac{q}{\epsilon_0} \tag{2.3}$$

where **da** is an element of area of the enclosing surface. Equation (2.3) is known as Gauss's law.

2.4 Electric potential

If a charge q_0 moves from point 1 to point 2 in an electric field **E**, the work done by the field is

$$W = q_0 \int_1^2 \mathbf{E} \cdot \mathbf{dl}$$

where **dl** is an element of length along a path from 1 to 2. The work done by the field is independent of that path from point 1 to point 2 and depends only on these two points. The electric field is said to be a *conservative* field. The difference in the potential energy between the final point 2 and the initial point 1, as the electric field moves the charge q_0 from 1 to 2 is

$$\Delta U = U_2 - U_1 = -W.$$

The negative sign indicates that, as the electric field does work on the charge to move it between those two points, the potential energy of the point charge decreases, being less at point 2 than at point 1.

The potential energy of q_0 depends on the nature of the electric field **E** and on the magnitude of the charge. It is then convenient to define the potential energy per unit charge, V, which depends only on the nature of the field, as follows,

$$\Delta V = V_2 - V_1 = -\frac{W}{q_0}.$$

The quantity V is called the *electric potential*.

If we chose point 1 to be the standard reference point which we can arbitrarily locate at infinity, and assign to the potential energy of q_0 when located at this point a value of zero, then V_1 is clearly zero and we can set V_2 to V. The potential at a point in an electric field is then,

$$V = -\int \mathbf{E} \cdot d\mathbf{l}. \tag{2.4}$$

2.5 Conductors in electrostatic fields

Conductors are a particular class of matter in which electric charges are free to move about *in* the material. In the case of solid metallic conductors, one or two electrons per atom are the charges that move about. In the case of liquid conductors, such as salty water, it is ions that are free to move in the liquid. Gases at extremely high pressures, such as hydrogen deep in the atmospheres of Saturn and Jupiter, are also liquid conductors. In this case, it is electrons freed from their hydrogen nuclei that are free to move.

Although conductors have free charges that can move about, an isolated conductor is still neutral. If an additional charge were to be placed inside an isolated conductor, the repulsive forces of the charges inside the conductor would push this additional charge out to the surface, where this mutual repulsion is balanced by surface forces. The time for this process to take place would be of the order of 10^{-14} s [2]. Thus, the net charge inside an isolated conductor is zero.

In electrostatics, there are no continuous sources of energy in a conductor and the electrons arrange themselves until the electric field everywhere inside the conductor is zero. This is true even if the conductor is placed in an external electric field. In this case, the external field exerts a force on the free electrons, moving them inside the conductor in an opposite direction to the field, leaving the positive ions in the metal unpaired. However, the conductor is still neutral. This separation of charges caused by the external field generates an internal electric field that opposes and cancels the external applied field, so that the net field inside remains at zero.

For the case of a charged spherical conductor, the charges are distributed uniformly over the surface of the sphere. The electric potential at the surface of a sphere of radius r holding a charge q is

$$V = Er = \frac{1}{4\pi\epsilon_0} \frac{q}{r}. \tag{2.5}$$

Using Gauss' law, it is not difficult to see that the electric field just outside a conductor is

$$E = \frac{\sigma}{\epsilon_0} \tag{2.6}$$

where σ is the surface charge density.

2.6 Capacitance

When charge is added to an isolated conductor, the charge stays on the surface of the conductor. The charge is now "stored" on the conductor. This charge can later be moved to another place. We can think of an isolated conductor as a place to store charge. If we have two isolated conductors nearby, we can use one as the source of charge and the other as the receptor by moving charges from one to the other. A separation of charges is produced, with one conductor holding with a charge $+q$ and the other a charge $-q$. This system is called a *capacitor*.

From Coulomb's law (2.1), we know that the electric field $\mathbf{E}$ that is set up between the two charged conductors is proportional to the charge q. The potential V between the two conductors is also proportional to q. The constant of proportionality is the *capacitance C* of the system of two conductors:

$$C = \frac{q}{V}. \tag{2.7}$$

Moving a charge (an electron) from one conductor to the other in a capacitor requires work. When the first electron is moved, a small electric field is set up across the two conductors. The electric field lines start at the positive ion left behind and end at the electron. Work must be done against this field to move the next electron across. This field clearly increases with the number of electrons transferred. If at a certain time a charge q has been transferred, the work needed to move the next dq is

$$dW = V dq = \frac{q}{C} dq.$$

The work needed to charge a capacitor to a total Q is

$$W = \int_0^Q \frac{q}{C} dq = \frac{1}{2} \frac{Q^2}{C} = \frac{1}{2} C V^2.$$

The charged capacitor has a final potential V across the conductors.

2.7 Electrostatic breakdown

The strength of the electric field around a conductor depends to some extent on the shape of the conductor. If the conductor has a sharp edge or point, for example, the field around the edge or point is much higher than on the other areas of the conductor.

Following Feynman [3], we can illustrate how the field is highest at the sharp edges by considering two charged conducting spheres of radii r and R (with r much smaller than R) connected by a thin wire (figure 2.2). If the large sphere has a charge Q and the small sphere holds a charge q, the electric potentials at the surface of each sphere are (see (2.5)):

$$V_R = \frac{1}{4\pi\epsilon_0} \frac{Q}{R} \quad \text{and} \quad V_r = \frac{1}{4\pi\epsilon_0} \frac{q}{r}.$$

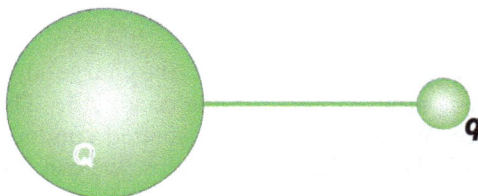

Figure 2.2. Two spheres of radii R and r holding charges Q and q, respectively, and connected by a thin wire can be used to illustrate how the field is highest at sharp edges.

But $V_R = V_r$, so that

$$\frac{Q}{R} = \frac{q}{r}.$$

Since the magnitude of the electric field at the surface of each sphere is proportional to the surface charge density (2.6),

$$\frac{E_R}{E_r} = \frac{\sigma_R/\epsilon_0}{\sigma_r/\epsilon_0} = \frac{Q/R^2}{q/r^2} = \frac{r}{R}.$$

Thus, the electric field is inversely proportional to the radius of curvature of the surface.

In planetary atmospheres, ions can be accelerated by the electric fields around conductors. If a conductor has sharp edges or points, the electric field can be large enough to accelerate ions to the speeds needed to ionize molecules, generating additional ions and electrons that can produce an avalanche leading to electrostatic breakdown in that atmosphere.

2.8 Dielectrics in electric fields

Dielectrics or insulators are materials in which all electrons are strongly bound to their atoms or molecules. Therefore, insulators have no conduction or free electrons. Since charge cannot flow freely in insulators, they have zero electrical conductivity.

An external electric field **E** can distort the charge distribution in an insulator, causing the electrons around a molecule to be microscopically displaced. The displacement can cause the molecule to stretch, in the case of non-polar molecules, or to rotate, in the case of polar molecules.

For a nonpolar molecule, the external field displaces the electrons away from the center of the molecule, in the opposite direction to the applied field **E**, and the nuclei of the atoms forming the molecule in the direction of the field. In this case, the centers of the positive charge and the center of the negative charge in the molecule are separated by a small distance, forming a small *dipole*. The molecule is said to be *polarized.*

When two opposite charges of equal magnitude q are separated by a distance s, it is convenient to define a vector quantity called the *dipole moment* **p** as

$$\mathbf{p} = q\,\mathbf{s}$$

where **s** is a vector of length s pointing away from the negative charge and toward the positive charge. For the case of an atom in an electric field **E**, the dipole moment can be written as

$$\mathbf{p} = \alpha \mathbf{E}$$

where α is the *atomic polarizability* of the atom.

In the case of polar molecules that have a permanent dipole moment built-in, such as the water molecule, the external electric field **E** will exert a torque on the molecule that will rotate the molecule to tend to align it with the field. For a molecule with a dipole moment **p**, the torque is

$$\tau = \mathbf{p} \times \mathbf{E}.$$

The polarization of insulators with nonpolar molecules and the alignment of the molecules in insulators with polar molecules disappears almost immediately after the external field is removed. This is due to the continuous interactions of the molecules in a solid. These interactions increase with temperature.

The polarization of the insulator which results from the alignment of all the dipoles can be represented by the total dipole moment per unit volume, **P**. This quantity is known as the *polarization*. In linear dielectrics, the polarization is proportional to the field,

$$\mathbf{P} = \varepsilon_0 \chi_e \mathbf{E}$$

where χ_e is the electric susceptibility of the medium.

When polarized by an external **E** field, an uncharged insulator remains neutral. Inside the insulator, the aligned dipoles form rows in which the positive end of a dipole is adjacent to the negative end of the next dipole. The opposite charges of equal magnitude from adjacent dipoles cancel each other out. Clearly the two ends of the row of dipoles are not paired up with opposite charges. The positive charge at the end is located near the insulator surface on the lee side of the field and the negative end on the side facing the field. These unpaired charges at the ends are called *bound* charges.

The build-up of opposite bound charges at each end of the dielectric by an external field **E** produces an internal electric field $\mathbf{E}_b$ in the opposite direction to that of **E**. This is the field due to the polarization of the dielectric. The resultant field $\mathbf{E}_d$ inside the dielectric is

$$\mathbf{E}_d = \mathbf{E} + \mathbf{E}_b.$$

$\mathbf{E}_d < \mathbf{E}$, since **E** and $\mathbf{E}_b$ are in opposite directions.

2.9 Plasmas

A highly ionized gas is a good conductor of electricity. The region in the gas with a balanced number of positive and negative charges is known as a *plasma*. When an ionized gas is placed in an external electric field, the ions and free electrons in the gas distribute themselves to shield most of the gas from the external field. The space charge at the boundary of the plasma is known as the *shield*.

If a small charge $+Q$ is placed in a plasma, electrons in the plasma are attracted to the charge while positive ions are repelled by the charge. The cloud of electrons surrounding the small positive charge Q shields the charge from other electrons, reducing their interaction. To an electron far away from Q, the electrons and positive ions balance out, so that the interaction with Q is the regular Coulomb interaction. An approaching high-energy ion can partially penetrate the cloud, leaving some of the shielding electrons behind. If the ion has an energy high enough to penetrate the cloud entirely, it will feel the full Coulomb repulsion.

In equilibrium, the electron density is

$$N_e = N_0 \, e^{\frac{U - U_0}{kT}}$$

where U is the potential of the electron in the field of the charge Q, U_0 is the plasma potential, N_0 is the electron density where $U = U_0$, T is the absolute temperature, and k is Boltzmann's constant [4]. A similar expression holds for the ion density. The potential U of a charge at a distance r from Q is given by

$$U = U_0 + \frac{Q}{4\pi\epsilon_0} e^{-\frac{r}{h}}$$

where h is the *Debye length* or shielding distance, given by

$$h = \sqrt{\frac{\epsilon_0 \, kT}{2N_0 e^2}} \, .$$

References

[1] Griffiths D J 2012 *Introduction to Electrodynamics* 4th edn (London: Pearson Education)
[2] Guru B S and Hiziroglu H R 1988 *Electromagnetic Field Theory Fundamentals* 2nd edn (Cambridge: Cambridge University Press)
[3] Feynman R P *et al* 1964 *Feynman Lectures on Physics* vol 2 (Reading, MA: Addison-Wesley)
[4] Reitz J R *et al* 2008 *Foundations of Electromagnetic Theory* 4th edn (Reading, MA: Addison-Wesley)

Chapter 3

Electrical breakdown and charge decay in planetary atmospheres

For the most part, electrostatic phenomena are determined by the physical and chemical properties of the different planetary atmospheres. The maximum amount of electrical charge that can be deposited on surfaces, the length of time it takes for these surfaces to discharge, the nature of this discharge, and other factors, depend on atmospheric properties.

In general, planetary atmospheres can be considered to be spherically symmetric. This is actually an approximation, since sunlight generates day–night and latitudinal modifications to an atmosphere [1]. An atmosphere has a vertical structure which depends mainly on its atmospheric pressure and temperature.

Atmospheric pressure is defined as the normal force exerted by atmospheric gas per unit area. This pressure changes with elevation. At any given height, the force is the weight of the atmospheric gas above that height. Atmospheric temperature changes in a more complex way. In the case of the terrestrial atmosphere, the temperature decreases with altitude in some regions of the atmosphere but increases with increasing altitude in other regions.

3.1 Electrical breakdown in planetary atmospheres

The strength of electric fields and the behavior of electrical charges either in the atmosphere or on the surfaces of materials depends mostly on the atmospheric properties that we have just discussed. The amount of charge that can accumulate on the surface of a material depends on the value of the atmospheric pressure, the physical properties of the material, and on the distance to other materials. Whether the charge remains on the surface or is transferred to nearby objects depends on the values of these parameters. When charge does transfer across a gap, electrical breakdown occurs.

doi:10.1088/978-1-6817-4477-3ch3

J S Townsend and collaborators studied the phenomenon of electrical breakdown in depth seventy years ago [2]. According to Townsend, if an electric field E is set up between two metal electrodes in a gas, any ions present in the gap between the electrodes will be accelerated in the field, ionizing some gas molecules and releasing electrons. In this case, N_0 electrons with mobility b will drift with a velocity u along the field lines. These electrons will collide with atoms in the gas. Depending on their kinetic energies, these electrons will ionize α atoms, releasing an equal number of electrons per unit path length, and will attach to η gas atoms per unit length. The coefficients α and η are the primary ionization and attachment coefficients. There are also secondary ionizations processes which include the release of electrons from the cathode by ion bombardment, the ionization of gas by positive ions and by photons, and photoemission by photons produced in the gas by electron ionization. A secondary ionization coefficient γ combines all these secondary processes. These three coefficients are known as the *Townsend coefficients* and are fairly well determined for many common gases.

If the potential across the electrodes in not too high, resulting in a relatively weak electric field, the number of electron–ion pairs that are formed decreases and the electrons and ions recombine. If the field is increased, the number of charges increases and, at some point, there is a transition to one form of sustained discharge. This transition constitutes a breakdown in the gas, called a spark, and usually occurs suddenly. For the spark discharge to be self-sustaining, there must be one secondary electron for every initial electron; that is, for a path length $\mathrm{d}l$,

$$\gamma e^{\alpha-\eta}\,\mathrm{d}l - 1 = 1. \tag{3.1}$$

The Townsend coefficients α, η, γ depend on the local electric field. This field is a function of the space charge density and is not uniform [3].

The breakdown potential depends on the product of the gap length d between electrodes and the gas number density. The relationship is known as Paschen's law [4]. Historically, however, instead of the gas number density, the pressure p of the gas is used in the expressions for Paschen's law. Figure 3.1 shows experimental values of the Paschen spark–breakdown potential as a function of pd for air and carbon dioxide.

Clearly, large values of pd are due to either higher pressures or large gaps. At large pressures, there are enough molecules per unit volume and electron collisions with these molecules are frequent, requiring larger potentials to accelerate the electrons so that they acquire kinetic energies large enough for ionization to occur. Similarly, if the gaps are large, there are enough molecules per unit volume even at moderate pressures for electron–molecule collisions to be frequent, thus requiring larger potentials for ionization. As either p or d decreases, the voltage required for ionization decreases to a minimum value.

At very low pressures, the gas density is low and collisions are rare. Therefore, larger potentials are possible before breakdown takes place. Likewise, with very small gaps, electrons are not able to reach the kinetic energies needed for ionization, and breakdown takes place at much higher potentials.

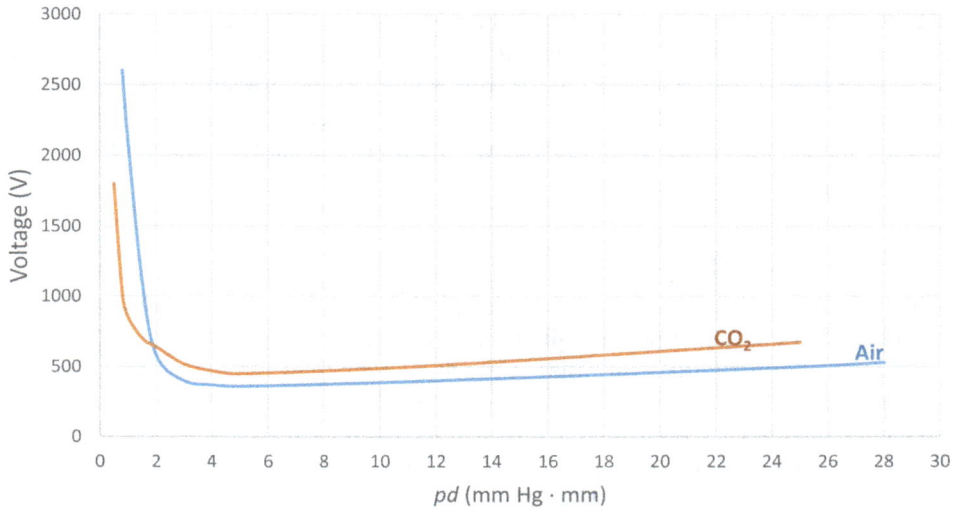

Figure 3.1. Paschen breakdown potential as a function of pressure–distance for air and carbon dioxide.

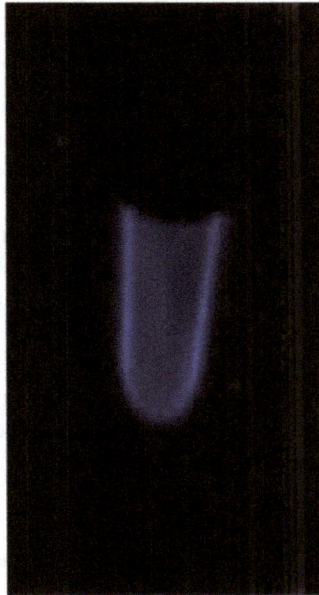

Figure 3.2. Corona in a region around a thin rod with a diameter of 0.64 cm at a 2.2. kV potential difference with a cylinder 9.6 cm in diameter. The corona current was 200 μA. (Courtesy of J S Clements, NASA.)

The data shown in figure 3.2 were obtained with clean, parallel plate electrodes without edge effects, so the field in the region between the electrodes at any point is never higher than at the center of the plane.

Electrostatic breakdown occurs naturally in planetary atmospheres. One important example is lightning. Although the processes that lead to lightning on Earth are not completely understood, the overall main process results from a large electrostatic potential difference between clouds and the ground, or between two clouds, that exceeds the breakdown threshold of the atmosphere at the pressure and conductivity conditions present. In addition to Earth, lightning occurs on Jupiter and Saturn and may occur on Venus. It is not known whether it occurs on Mars or on Titan, Saturn's largest moon.

3.2 Glow discharges and ion wind

A glow discharge, also known as a *corona* discharge, takes place at a lower voltage than a spark discharge. When a high voltage is set up between two electrodes, one of which has a radius of curvature that is much smaller than the gap, a strongly non-uniform electric field is formed in the region. In the vicinity of the sharp electrode, the electric field is very high. A free electron in this region will accelerate, knocking off electrons from neutral atoms and creating electron–ion pairs. The newly freed electron is also accelerated in the strong field, producing additional electron–ion pairs. In this way, an electron avalanche is formed in the region close to the sharp electrode. The ionization process emits electromagnetic radiation. In the atmosphere of the Earth, this radiation is in the blue region of the electromagnetic spectrum. The shape of this glow is in the form of a tall crown, giving it its name of *corona* discharge (figure 3.2).

If the voltage between the electrodes is not increased, any electrons in regions farther away from the sharp electrode do not gain enough kinetic energy to ionize the gas atoms and the discharge remains localized and stable. There is no breakdown across the electrodes.

When the sharp electrode is negative (*negative corona*), the electrons outside the high-field region, without enough energy to release other electrons from neutral atoms, attach themselves to some of these atoms, forming negative ions. These ions travel at low speeds toward the positive electrode and constitute the only current outside the high-field region. There is no glow in this external region. The positive ions formed by electron collisions with neutral atoms in the high field region accelerate toward the sharp, negative electrode, where they release secondary electrons that contribute to the discharge mechanism, as happens in the electrical breakdown case.

When the sharp electrode is positive (*positive corona*), the secondary mechanism is less important. The electrons created by ionization in the high-field region accelerate toward the positive sharp electrode while the positive ions move toward the negative electrode.

In both cases, the slow moving ions outside the high-field region, which have the polarity of the sharp electrode, collide with the atmospheric gas molecules, setting them in motion. The gas molecules move away from the sharp electrode, forming what is known as an *ion wind*. In the Earth's atmosphere, the ion wind speeds are about 1 m/s.

The voltage at which a stable corona is established depends on the gas and the electrode geometry. With very small and smooth electrodes, slowly increasing the

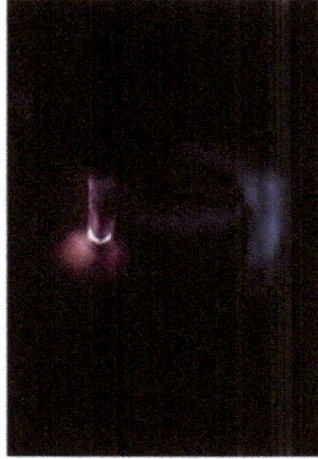

Figure 3.3. The pink glow of a streamer corona in air. (Courtesy of J S Clements, NASA.)

voltage above the corona threshold results in an increase in the corona current that is proportional to the square of the applied voltage. This enhanced corona is called an *ultracorona*.

With larger electrodes, an increase in the corona voltage breaks up the corona glow into separate filaments or streamers (figure 3.3). This *streamer corona* is still a stable glow discharge, with the current being carried by the ion flux. Increasing the voltage further will increase the length of the streamers until they reach the opposite electrode, at which point breakdown will occur.

Corona discharges occur naturally on Earth and may occur on many other planets, such as Jupiter, Saturn, and Venus. They may also occur on Mars and Titan.

3.3 Charge mobility

Electrical breakdown and glow discharges in planetary atmospheres are affected by the conductivity of the atmosphere, the presence of ions, and the ability of those ions and of charged particles to move. *Charge mobility* is a measure of the ability of an electric charge to move in an atmosphere while in the presence of an electric field.

The equation of motion for a charged particle in an external electric field E is

$$F = \frac{dp}{dt} = qE$$

where p is the particle's momentum, q is the particle's charge, and F is the electric force. If the particle moves with an average velocity v, the force imparted on the particle during a time t_c between collisions, the equation of motion can be written as

$$\frac{mv}{t_c} = qE$$

where m is the particle's mass. The average time between collisions is called the *mean free path*. The particle's average velocity is then

$$v = \frac{qt_c E}{m} = \mu E \tag{3.2}$$

where μ, written as

$$\mu = \frac{qt_c}{m} \tag{3.3}$$

is the *charge mobility* of the particle. If the particle has a fixed charge and mass, such as an electron of charge e and mass m_e, and moves in an atmosphere with a given mean free path, the charge mobility is constant. Thus the average electron velocity is proportional to the electric field and the constant of proportionality is the electron charge mobility.

The electron mean free path in the Earth's atmosphere near the surface is 6.63×10^{-8} m [5]. On Mars, with an atmospheric pressure about one hundredth of the Earth's, the mean free path near the surface is roughly 100 times longer or 4.8×10^{-6} m.

3.4 Charge decay in planetary atmospheres

Electric charges at rest on the surface of an isolated object will remain on that surface for a length of time that is mostly determined by the physical properties of the atmosphere in which the object resides. Charges need a conductive path to leave the surface. This conductive path is usually provided by ions present in the atmosphere. Cosmic rays and similar types of ionizing radiation reaching the atmosphere in the vicinity of the object generate ions as they interact with the neutral ions. In the absence of an electric field, these ions are distributed more or less uniformly in the volume around the object. The process generates positive and negative ions. When charge is present on the surface of the object, the electric field produced by this charge accelerates these atmospheric ions, thus generating an electric current in the atmosphere.

If the charged object is a conductor with surface charge density σ, the electric field E around the conductor is

$$E = \frac{\sigma}{\epsilon_0}. \tag{3.4}$$

The ions present in this electric field are accelerated by the field, generating an electric current j given by

$$j = \frac{E}{\rho} \tag{3.5}$$

where ρ is the electrical resistivity of the atmosphere in this region. Both positive and negative ion currents will be present, with the current of the same polarity as that of the charge on the conductor flowing away from the conductor and the current of

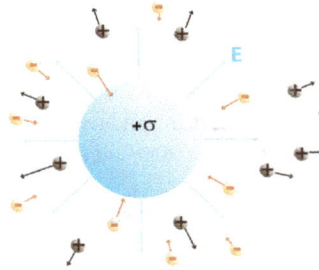

Figure 3.4. Ion currents in the vicinity of a charged conductor.

opposite polarity flowing toward the conductor (figure 3.4). The ion current flowing toward the conductor will clearly cause the charge density of the conductor to decrease at a rate $d\sigma/dt$, given by (3.4) and (3.5):

$$j = \frac{d\sigma}{dt} = \frac{1}{\epsilon_0 \rho}\sigma. \tag{3.6}$$

Since ρ is constant in the vicinity of the conductor, the solution to this differential equation is

$$\sigma = \sigma_0 \, e^{-t/\epsilon_0 \rho}$$

or

$$\sigma = \sigma_0 \, e^{-t/\tau} \tag{3.7}$$

where $\tau = \epsilon_0\rho$ is the *time constant* and σ_0 is the initial charge density of the conductor. As (3.7) indicates, charge decay in a conductor is exponential. The time constant τ is the time it takes the charge to decay to $1/e$ of its initial value. The time constant depends on the properties of the molecules composing the atmosphere and on the number of ions present.

Charges on insulators are not free to move and the mechanism of charge decay is different. Depending on the nature of the insulator, the condition of its surface and of any gases adsorbed, as well as on the conductivity of the atmosphere, charge may move through the insulator and/or across its surface [6]. Surface charge on an insulator can also be neutralized by ions in the atmosphere. As was the case with the charged conductor, an electric field arises due to the surface charge on the insulator. Because the charge on the insulator is not necessarily uniformly distributed around the surface, the electric field lines are not always perpendicular to the surface. At the edges of a charged patch on the surface, the field lines will point away from the normal and may even bend to fall on an area of the surface that may have a charge of opposite polarity. Ion currents will flow along those field lines bringing ions of opposite polarity that will neutralize some of the charge on the insulator surface. Charge decay in insulators does not follow the exponential decay form of (3.7). This more complex decay follows a modified hyperbolic law [7].

References

[1] Chamberlain J W and Hunten D M 1987 *Theory of Planetary Atmospheres* (San Diego: Academic) pp 3–7

[2] Townsend J S 1947 *Electricity in Gases* (Cambridge: Cambridge University Press)

[3] Cross J 1987 *Electrostatics: Principles and Applications* (Bristol: Adam Higler)

[4] Paschen F 1889 *Wied. Ann.* **37** 69

[5] National Oceanic and Atmospheric Administration National Aeronautics and Space Administration and U S Air Force 1976 *US Standard Atmosphere* (Washington, DC: US Government Printing Office)

[6] Kinderberger J and Lederle C 2008 Surface charge decay on insulators in air and sulfurhexaflorid—Part II: Measurements *IEEE Trans. Dielectr. Electr. Insul.* **15** 949–57

[7] Seaver A 2002 An equation for charge decay valid in both conductors and insulators *ESA-IEJ Joint Meeting (Northwestern University, Chicago)*

Chapter 4

The terrestrial electrostatic environment

4.1 The Earth's atmosphere

The atmosphere of the Earth is a mixture of many gases. Close to the surface, the atmosphere is composed of 78% nitrogen, 21% oxygen, and traces of many other gases, including water vapor, argon, and carbon dioxide. This composition changes with altitude. The different variations of temperature with altitude are due to the differences in chemical composition in each region of the atmosphere which, in turn, affect the way each region is heated. It is this variation in temperature that divides the atmosphere into distinct regions or layers.

The layer closest to the surface, below 12 km, is called the *troposphere*. The temperature of the atmosphere in this layer decreases with altitude. The lower end of the troposphere is heated by the Earth's surface. The upper end, away from the surface, is cooler (figure 4.1). The convection currents that result from this temperature difference are the main cause of terrestrial weather.

Right above it, reaching to an altitude of 40–50 km, lies the *stratosphere*. Part of the oxygen in this layer of the atmosphere is in the form of ozone, O_3, which absorbs ultraviolet (UV) and infrared radiation from the Sun. This direct absorption of the Sun's energy causes a temperature inversion in the stratosphere, with temperature rising with increasing altitude.

The next layer, extending to about 80 km from the surface, is the *mesosphere*. This layer does not contain ozone to absorb the Sun's energy. Atomic oxygen, O, and carbon dioxide radiate energy to space, and the temperature decreases with altitude, reaching a value of about 190 K at the top of the mesosphere.

In the *thermosphere*, which starts at an altitude of about 80 km, temperature again increases with increasing altitude. The extremely low density of this region prevents the formation of oxygen and nitrogen molecules (O_2 and N_2) that are found close to the surface. The atmosphere is heated directly, as atomic oxygen and nitrogen absorb UV radiation from the Sun. The temperature in the thermosphere increases

doi:10.1088/978-1-6817-4477-3ch4   4-1

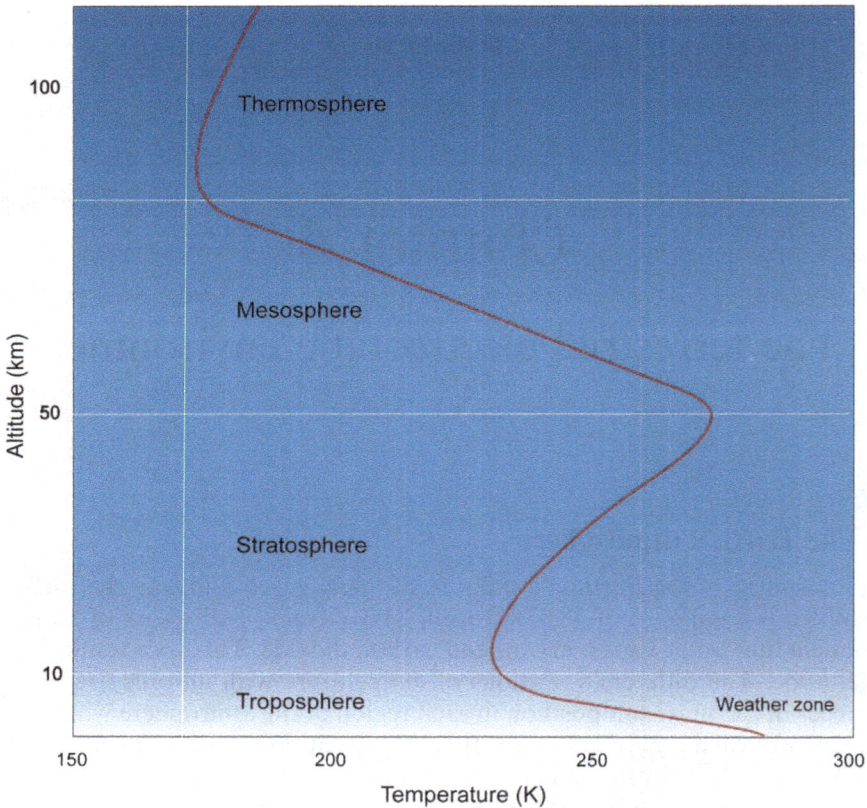

Figure 4.1. Temperature variation (red line) in the Earth's atmosphere.

with altitude, since the most energetic photons from the Sun are absorbed in the upper layers of the atmosphere.

The temperature of the atmosphere at an altitude of 400 km, where the International Space Station (ISS) orbits, is above 750 K. However, the density of the atmosphere is very low in this layer, with too few fast-moving atoms to cause an increase in temperature on the ISS.

The *ionosphere*, which starts in the mesosphere, is the region starting at about 50–70 km and reaching to about 2000 km above the surface where atmospheric atoms are ionized by solar radiation and cosmic rays. The electrons knocked off from the nitrogen and oxygen atoms in the atmosphere behave as free particles. The highest electron densities exist in the region between 200 km and 400 km. This is the *F* region and it is ionized by the radiation from the Sun during the day and by cosmic rays at night.

The pressure in the atmosphere decreases steadily with altitude throughout these layers. At sea level, the average atmospheric pressure is 1 *atmosphere* (atm), equal to 101 325 N/m^2 or 1.013×10^5 Pa (= 1.013 bar = 1013 millibar = 14.70 lb/in^2). As we go up in the atmosphere, there is less air above us and the pressure decreases. Table 4.1 shows the values of atmospheric pressure at different altitudes.

Table 4.1. United States Standard values of atmospheric pressure at different altitudes.

Altitude	Atmospheric pressure			
Meters	mbar	kPa	mm Hg	kPa (From barometric formula)
0	1.013	101.33	760	101.33
1000	0.899	89.88	674	90.00
2000	0.795	79.50	596	79.94
3000	0.701	70.12	526	71.01
4000	0.617	61.66	462	63.07
5000	0.541	54.05	405	56.02
6000	0.472	47.22	354	49.76
7000	0.411	41.11	308	44.20
8000	0.357	35.65	267	39.26
9000	0.308	30.80	231	34.87
10 000	0.265	26.50	199	30.98
15 000	0.121	12.11	90.8	17.13
20 000	0.0553	5.529	41.5	9.47
25 000	0.0255	2.549	19.1	5.24
30 000	0.0120	1.197	8.98	2.90
40 000	0.00287	0.287	2.15	0.89
50 000	0.00080	0.080	0.60	0.27
60 000	0.00022	0.022	0.16	0.0827
70 000	0.00005	0.005	0.04	0.0253
80 000	0.000011	0.001	0.01	0.0077

4.2 Electrical breakdown in the terrestrial atmosphere

As stated in chapter 3, electrical breakdown potentials as a function of pressure and electrode gap (Paschen's law) are obtained experimentally. These electrodes are usually parallel plate electrodes with rounded edges that have a radius of curvature decreasing gradually so that the field has a maximum value at the center of the plane region. Figure 4.2 shows experimental values of the Paschen breakdown potential as a function of pressure-distance for air.

Figure 4.3 shows experimental values of the breakdown potentials in air near the surface of the Earth for different parallel-plate electrode gaps. With a 1 cm gap, the breakdown potential is about 30 kV, a value often quoted. However, with a 2 cm gap the potential is 57 kV and with a 4 cm gap it is 109 kV, not exactly doubling or quadrupling the value for the 1 cm gap. Figure 4.4 shows how the breakdown potential decreases rapidly as atmospheric pressure decreases with altitude.

4.3 Radiation from the Sun: the solar wind

The gas in the Sun's corona has temperatures above 10^6 K, hot enough for particles to escape the Sun's gravity, moving outward at speeds of about 400 km/s. This speed

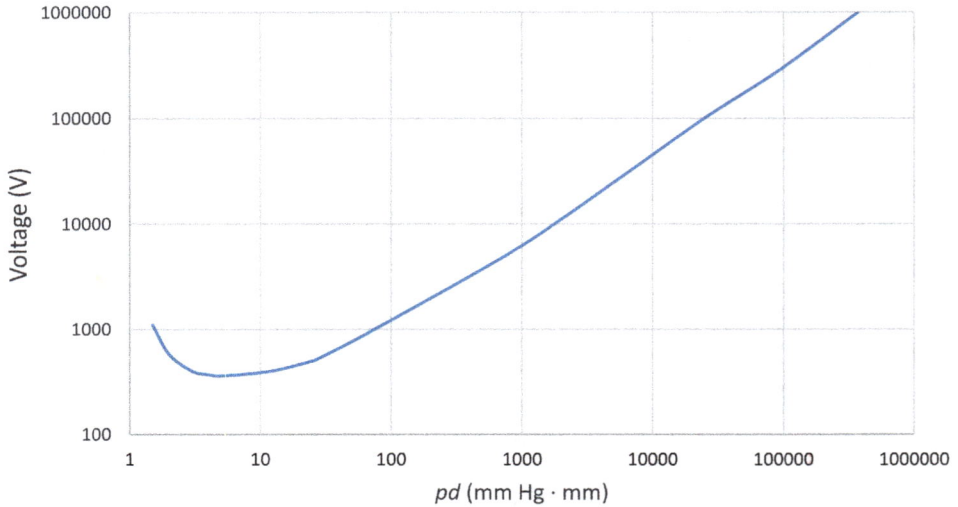

Figure 4.2. Paschen breakdown potentials for parallel-plate electrodes as a function of pressure–distance.

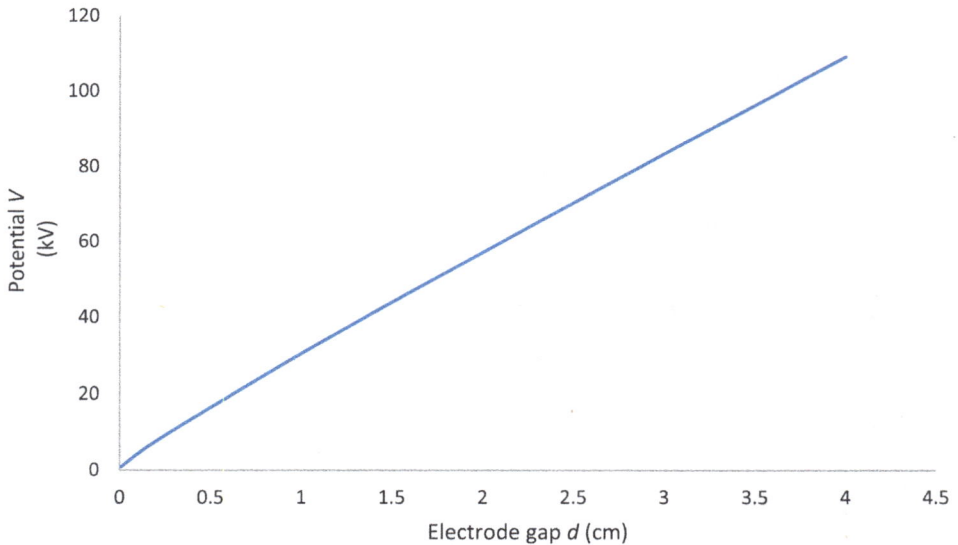

Figure 4.3. Breakdown potentials for several parallel-plate electrode gaps near the surface of the Earth, at 1 atm.

makes the solar wind supersonic, since it is faster than the speed of sound in the rarefied gas that exists between the planets. Despite carrying about a million tons of solar matter each second away from the Sun, the particle density in the solar wind in the vicinity of the Earth is very thin, since it drops off as r^{-2}. The energy density of the solar wind at 1 AU is about 5×10^3 eV/cm^3 [1].

The charged solar wind particles are mostly electrons and protons with some helium nuclei and traces (about 0.1%) of heavier ions. The solar wind pushes the tails of comets away from the Sun and shapes the magnetic fields around the planets.

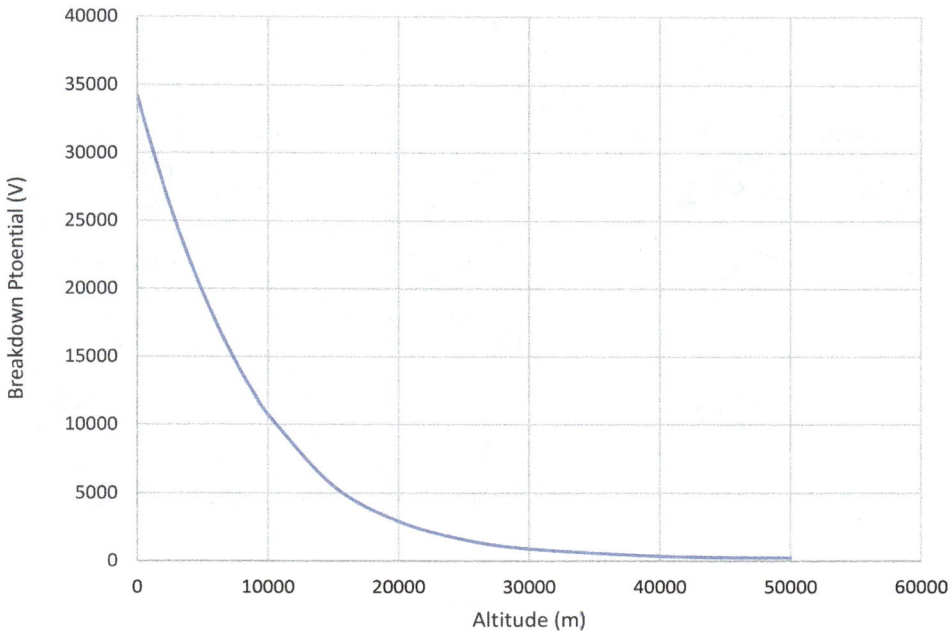

Figure 4.4. Breakdown voltage for different altitudes for a 1 cm gap.

4.4 Radiation belts

The interaction of the Earth's magnetic field with the solar wind produces a semipermeable barrier to the charged particles in the solar wind called the *magnetosphere*. Most of the solar wind particles flow around the outer edge of the magnetosphere. As they encounter the magnetosphere, the supersonic charged particles in the solar wind form a shock wave (figure 4.5) that abruptly slows down the supersonic charged particles to subsonic speeds. The front of the bow shock is located at about ten Earth radii, but its exact location depends on changes in solar activity. The magnetosphere stretches out away from the Sun to about 60 Earth radii, forming a long tail that reaches beyond the orbit of the Moon.

When a charged particle from the solar wind enters the Earth's magnetosphere with a velocity **v** at right angles to the magnetic field **B**, the magnetic force acting on the particle will be perpendicular to both the particle's velocity and the magnetic field. This magnetic force is

$$\mathbf{F} = q\mathbf{v} \times \mathbf{B}$$

where q is the particle charge. As the vector product between **v** and **B** indicates, this force accelerates the particle in a direction that is perpendicular to its velocity, making the particle move in a circle at a constant speed. If the charge particle enters the magnetosphere with a velocity component that is parallel to the field **B**, the particle will move along the field but the perpendicular component of its velocity will also force it to move along a circle. The particle will then follow a spiral path.

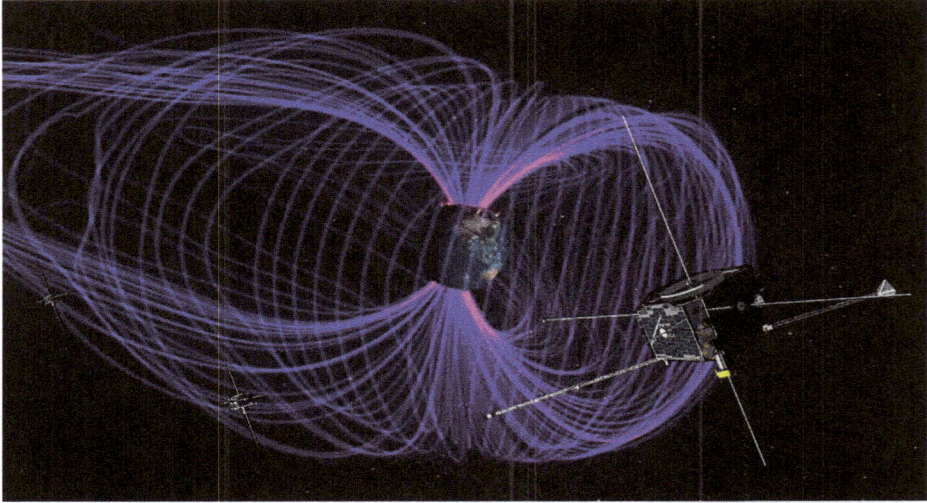

Figure 4.5. The solar wind interacts with the Earth's magnetic field forming a shock wave that slows down the particles. NASA's THEMIS mission has helped map out the magnetosphere with five nearly identical satellites launched since 2007 to orbit the Earth in highly elliptical orbits. (Courtesy of NASA.)

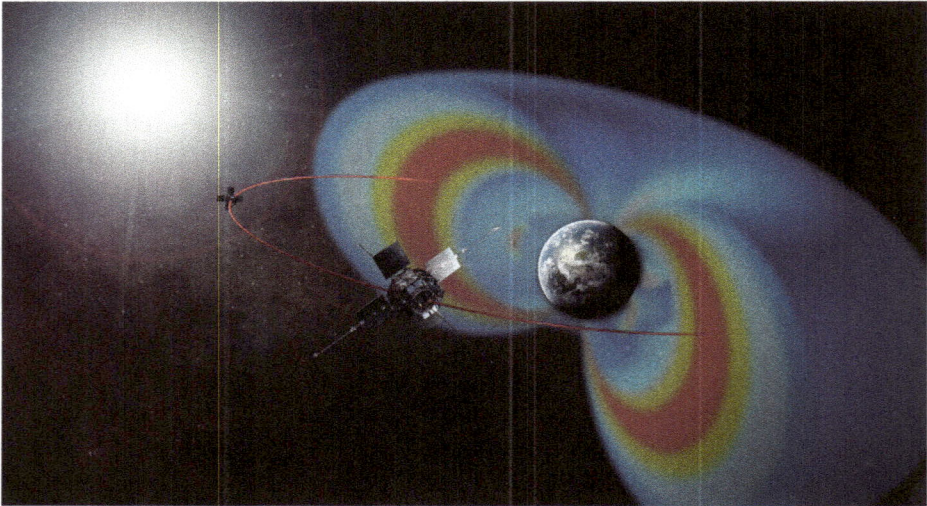

Figure 4.6. The Van Allen radiation belts surrounding the Earth. NASA's Van Allen Probes, launched in 2012, orbit through the belts to study them. (Courtesy of NASA.)

But, since the Earth's magnetic field is stronger near the poles, the particle will oscillate back and forth in a spiral pattern, trapped in the magnetic field. There are two doughnut-shaped regions of trapped charged particles from the solar wind around the Earth. These are the *Van Allen belts* (figure 4.6). The inner Van Allen belt, extending from about 2000–5000 km, traps mostly protons. The outer belt, at

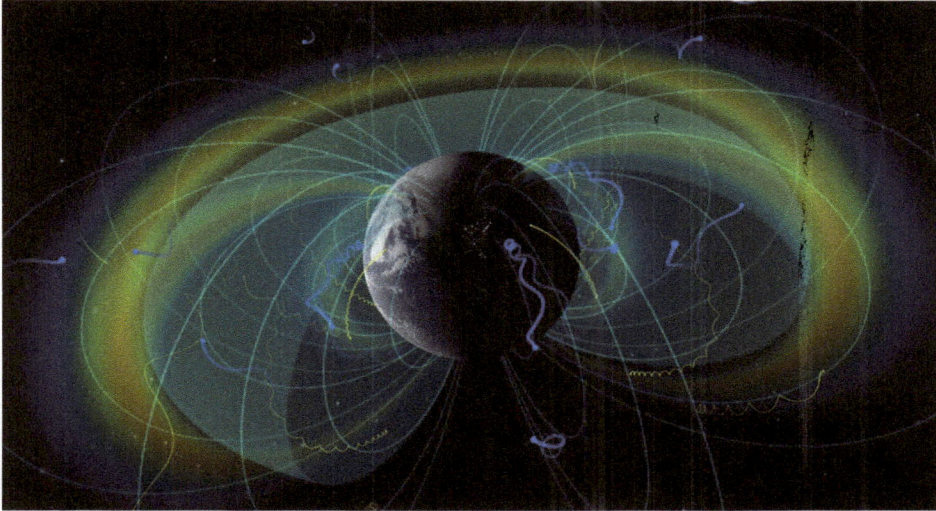

Figure 4.7. The inner edge of the outer radiation belt (shown in yellow) is a hard boundary that stops the most energetic electrons from entering the gap between the radiation belts. The plasmasphere, shown in green, is sliced open to reveal the inner radiation ring. (Courtesy of NASA.)

altitudes between 13 000 km and 19 000 km, traps mostly electrons. Electric and magnetic fields in this belt speed up electrons to nearly the speed of light [2]. The inner edge of the outer belt is a sharp boundary that prevents the most energetic electrons from reaching Earth [3].

Data from NASA's Van Allen Probes mission, launched in August of 2012 to study the radiation belt region (figure 4.6), has shown the cause for the existence of the hard boundary and of the existence of two radiation belts rather than a single belt. Electrons in the outer belt move at high velocities around the Earth, with only a small velocity component toward the Earth. Other charged particles in the *plasmasphere*—a relatively cool region that starts at about 1000 km and extends into the outer radiation belt—interact with the less energetic electrons at the inner edge of the outer belt causing them to scatter. These scattered electrons repel the drifting high-energy electrons away from the Earth, keeping them out of the gap between the two radiation belts (figure 4.7).

4.5 Auroras

Violent and sudden eruptions in the solar atmosphere, called *solar flares*, raise the temperature in the region to 5×10^6 K, emitting vast amounts of particles and radiation at all wavelengths. *Coronal mass ejections*, lasting several hours, are even more violent, sending more than 10^{12} kg of energetic plasma and radiation into space.

If a solar flare or a coronal mass ejection happens to be directed toward the Earth, the ejected particles interact with the Earth's magnetic field generating magnetic

Figure 4.8. Aurora photographed by one of the astronauts of Expedition 6 aboard the ISS on March 30, 2003. (Courtesy of NASA.)

storms that dump some of the charged particles from the radiation belts into the atmosphere. Atoms of nitrogen and oxygen absorb the UV radiation that results from the collisions of these fast charged particles with the atoms in the upper atmosphere. These nitrogen and oxygen atoms then emit this energy as visible light, generating the *auroras* (*aurora borealis* or *northern lights* and *aurora australis* or *southern lights*). Figure 4.8 shows an aurora photographed from the ISS.

Auroras are a very dynamic phenomenon, undergoing large-scale changes, from being relatively static to having a very active condition several times each night. These abrupt changes are known to be triggered by *substorms*, and cause the auroras to flicker, sending ions toward the Earth. The existence of these substorms has been known for some time, but their cause has been a mystery. A few years ago, NASA's THEMIS spacecraft fleet, launched in 2007, discovered that particles in the solar wind actually flow along giant *magnetic ropes*, twisted bundles of magnetic field lines that directly connect the Earth's atmosphere to the Sun. More recently, THEMIS researchers proposed that substorms may be caused by bubbles of low-density plasma that ride along these magnetic ropes toward the Earth [4].

References

[1] Lewis J S 1995 *Physics and Chemistry of the Solar System* (San Diego, CA: Academic)
[2] Kaneka S G *et al* 2015 Inner-calibration energetic electrons and proton measurements by MagEIS, RRPT and RPS instruments onboard Van Allen Probes *Conf. Measurement Techniques in Solar and Space Physics (Boulder, CO)*

[3] Baker D N *et al* 2014 An impenetrable barrier to ultrarelativistic electrons in the Van Allen radiation belts *Nature* **515** 531–4

[4] Pritchett P, Coroniti F B and Nishimura Y 2014 The kinetic ballooning/interchange instability as a source of dipolarization fronts and auroral streamers *J. Geophys. Res. Space Phys* **119** 4723–39

Chapter 5

Spacecraft and satellites in the electrostatic environment of the Earth

5.1 Spacecraft and satellite orbits

Spacecraft and satellites orbiting the Earth are exposed to the electrostatic environments of the Earth's atmosphere. There are three main orbital regions around the Earth: *low Earth orbit* (LEO), *medium Earth orbit* (MEO), and *geosynchronous Earth orbit* (GEO).

LEO is the region with an altitude between 160 km and 2000 km. Most scientific satellites, many weather satellites, and the ISS are in LEO. The Hubble Space Telescope also orbits in LEO. One special kind of LEO is the *polar orbit*, where satellites orbit the Earth form pole to pole, taking about 99 min to complete an orbit. A satellite in polar orbit will cross from daylight to night-time as it crosses the pole and will view most of the Earth in one day, once at night and once in daylight.

MEOs lie in the space below GEOs (which are at an altitude of 35 786 km) and above 2000 km. These orbits are used for communications, navigation, and science satellites. The Global Positioning System (GPS) satellites fly on near-circular MEOs at an altitude of 20 200 km, which gives them an orbital period of 12 h. Since a satellite on this orbit will cross over the same spot on the Earth twice in a day, the orbit is called semi-synchronous.

Satellites in GEOs, which have a radius of 42 164 km, orbit at the same rotational speed as that of the Earth. If a GEO is circular with an inclination of zero, the orbit is called *geostationary*. A satellite on a geostationary orbit stays directly above the same spot on the Earth's surface. Geostationary orbits are useful for weather and communications satellites.

5.2 Spacecraft charging

Spacecraft and satellites in any of these orbits may experience changes in the electrostatic potential at their surfaces relative to the surrounding plasma. Spacecraft

doi:10.1088/978-1-6817-4477-3ch5

charging depends not only on the spacecraft's shape and composition but also on its orbit. Spacecraft surfaces facing the Sun can acquire a net positive electrostatic charge with the release of electrons by photoemission due to UV radiation from the Sun. Since solar UV photons will not interact with the surfaces of the satellite in the shade, differential charging between the Sunlit and shadow regions will occur. This differential charging is larger on insulating surfaces.

Charged particles in the plasma from the different regions crossed by a satellite in orbit can also charge the satellite. The amount of charge depends on the energies of the plasma particles in the region, which range from electronvolt to kilo-electronvolt levels. The 'wishbone chart' from the *NASA Technical Handbook* shown in figure 5.1 illustrates the calculated electrostatic surface charging expected for a satellite orbiting at altitudes from 100–100 000 km for some latitudes along with the hazard level [1]. The chart does not include calculations for polar Earth orbits. These calculations were performed for aluminum spherical satellites in the dark (to prevent photoelectric charging). At latitudes between 40 and 50 degrees north, the potentials reach values of over 400 V, which may produce discharges. The charging asymmetry for the northern and southern latitudes is due to the tilting of the magnetic north pole relative to the geographic pole. The values of the electrostatic potentials on this chart are to be taken as the worst-case surface charging for a spacecraft in the Earth's electrostatic environment [1].

As figure 5.1 shows, regions where the electrostatic potentials reach levels that may cause spacecraft charging vary with latitude and altitude. Spacecraft in GEO at certain latitudes are particularly vulnerable, with potentials reaching over 28 000 V.

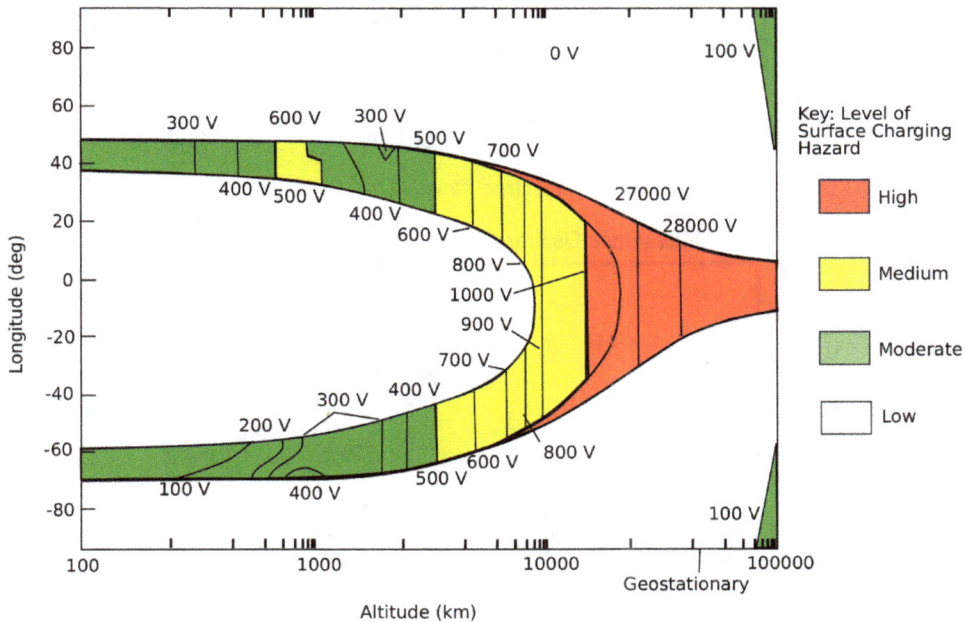

Figure 5.1. Calculated values of the charging potentials for spherical metallic satellites on orbits at some latitudes for altitudes between 100 and 100 000 km. (Courtesy of NASA.)

Spacecraft launched from the surface of the Earth are also exposed to the different electrostatic environments as they fly to their orbits through different regions of the atmosphere.

5.3 Spacecraft charging in LEO

LEO lies in the ionosphere. In this region, at altitudes above 90 km, UV radiation from the Sun splits diatomic oxygen molecules into individual atoms. Solar radiation and cosmic rays ionize the oxygen and nitrogen atoms, producing free electrons and ions with equal densities. The concentrations of these electrons and ions are relatively high and change with altitude and solar activity [2]. Since electrons have higher mobilities, they are collected more easily, negatively charging spacecraft in this region. Although the LEO plasma density is high, the energy is normally low, at about 0.1 eV. Thus spacecraft charge only to a few volts negative, which is not an issue [3]. Figure 5.2 shows the electron density throughout most of the ionosphere for periods with minimum and maximum solar activity [4]. The highest electron concentration, at about 300 km, lies in the *F* region of

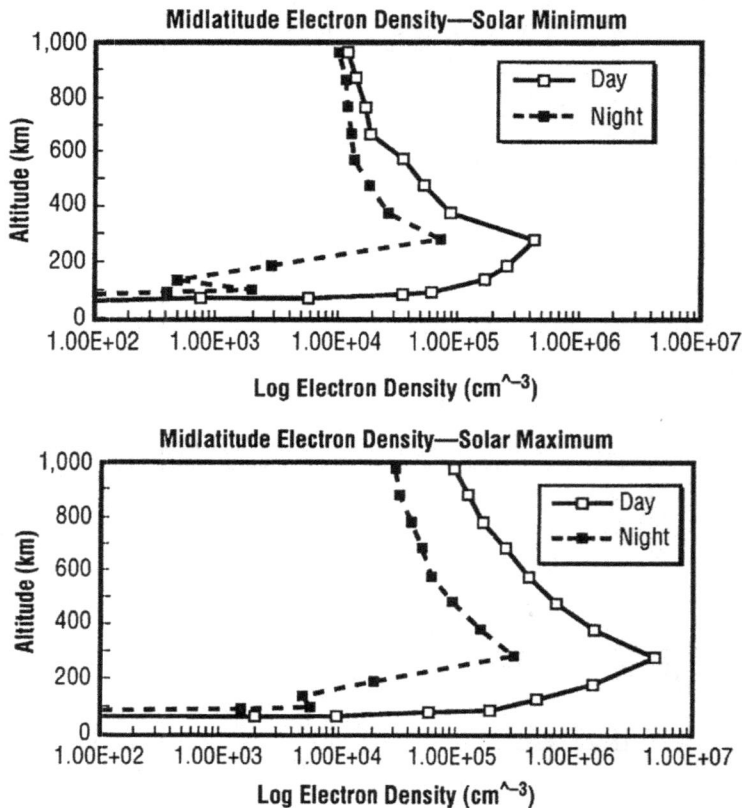

Figure 5.2. Variation in electron density at different altitudes in the ionosphere for a solar minimum (top) and solar maximum (bottom) [4]. (Courtesy of NASA.)

the ionosphere. This region, extending from 200–400 km, is where the ISS, the Hubble Space Telescope, and most scientific satellites orbit.

Orbiting spacecraft in the F region move at about 8 km/s in the dense, low-energy plasma. This orbital velocity is slower that the electron thermal velocities, which are about 200 km s^{-1}, but faster than the 1 km s^{-1} thermal velocities of positive ions. Thus, spacecraft in LEO orbit supersonically relative to the ambient ions and subsonically relative to the thermal electrons. As a result, a wake forms behind the spacecraft where the ion density is much smaller than that of the ions in the plasma. This unequal flux results in electrons impacting all LEO spacecraft surfaces while ions can only impact ram surfaces [2, 3]. However, electron impact is not uniform throughout the spacecraft. The faster moving electrons that enter the wake form a space charge that repels additional electrons, excluding them from the wake region [1, 5].

Spacecraft in polar orbits, even at altitudes of about 300 km, can charge up to several kilovolts due to high-energy electrons reaching those low altitudes. Since the geomagnetic field lines plunge at polar orbits, charging conditions at these altitudes are similar to those in GEO, where spacecraft can charge to potentials in the kilovolt range.

5.4 Charging of the ISS

The ISS orbits the Earth in the F region of the ionosphere, at an altitude of 400 km at an inclination of 51.6 degrees. As stated earlier, the relatively high plasma density in LEO limits spacecraft surface charging to a few volts negative. Due to its large size and power requirements, and the inclination of its orbit, the ISS has three additional sources of charge that can be important: the large photovoltaic array system, the magnetic induction process, and the interaction with the high-energy electrons in the polar regions.

Electric power systems on spacecraft normally operate at the 28 V DC standard of the aircraft industry. Interaction with the ionospheric plasma at this relatively low voltage generates low surface charges and is not a concern. However, ISS is not just a spacecraft but a habitat, with much larger power requirements. The ISS photovoltaic power system operates at 160 V to reduce weight and power loss. As is the case for most spacecraft, the solar array system on ISS has the negative end connected to the ISS ground. This set up is due to the scarcity of space qualified power systems with a positive ground when ISS was being constructed. With this arrangement, the positive end of the arrays collects electrons from the dense ionospheric plasma while the negative end collects ions. Although the ion and electron currents are about the same, the electron current density is much higher, and the array and consequently ISS charge negative to values that can reach about –80 V [6]. On their own, these array voltages need to be mitigated perhaps by encapsulation of the array edges to prevent arcing due to contamination of the arrays or by bakeout in orbit at 100 °C for about a week to remove contamination [2].

The second source of charge on ISS is the magnetic induction process that results from the motion of the conductive components of ISS through the magnetic field of the Earth. From Faraday's law we know that a magnetic field **B** produces an electromotive force (EMF) that drives a current through a conductor moving with a velocity **v** in that magnetic field. The potential difference of this *motional* EMF is given by

$$\varepsilon_{\mathrm{m}} = \oint \mathbf{v} \times \mathbf{B} \cdot \mathbf{dl}$$

where **dl** is an element of length in the conductor pointing in the direction of the current. Long, conducting ISS surfaces moving with a velocity component perpendicular to the Earth's magnetic field can generate these voltages. The effect is larger when ISS orbits through the more concentrated magnetic field lines near the Earth's poles. Potentials of about −90 V relative to the plasma environment have been measured on the photovoltaic arrays [2, 6, 7]. These values are not of concern by themselves but, when combined with the negative photovoltaic charging, they can reach values that need to be mitigated.

The third source of ISS charging is due to the daily passage of ISS through the auroral or polar region, where the Earth's magnetic field lines plunge, allowing for high-energy electrons in the 7–10 keV range to reach LEO altitudes. Measurements of the satellites of the Defense Meteorological Satellite Program, which orbit in the auroral region at an altitude of 840 km, show that they can charge to −2 kV when they encounter intense electron flux [3]. At the lower altitude at which ISS orbits, the plasma density is much higher and high-level charging should not normally take place. However, the plasma density in the wake can be up to two orders of magnitude lower [3]. High-energy electrons in the auroral regions can charge surfaces in this area. Data from the US Department of Defense/European Space Agency/Russian Roscosmos State Corporation satellite, suggest that the probability for this charging effect is low but not zero [7].

To mitigate these problems, NASA installed a system to ground ISS to the ambient plasma. This system, called the *plasma contactor unit* (PCU), generates a high-density plasma that makes electrical contact with the ambient plasma [8]. The system uses a cathode that converts a small supply of xenon gas into ions and electrons that are discharged to space, thus carrying away the excess charge (figure 5.3). With the PCU in place, voltages on ISS conducting surfaces do not exceed ±20 V [2]. For redundancy, two PCUs were installed on the Z1 truss of ISS. The units operate only during astronaut extra-vehicular activities and during spacecraft docking.

Insulator surfaces on the ISS in general do not build up charges due to the neutralizing effect of the surrounding plasma. This neutralizing effect does not happen in the wake, where, as stated above, the plasma density can be two orders of magnitude lower. In the unprotected region of the wake, differential charging can develop on insulating surfaces when the ISS passes through the auroral region, where it is exposed to high-energy electrons.

On 11 May 2009, NASA launched the Space Shuttle to repair and replace several instruments aboard the Hubble Space Telescope, which orbits in LEO. One of the repair tasks was to replace the low-voltage power supply board for the Space Telescope Imaging Spectrograph. This task required the astronauts to expose the

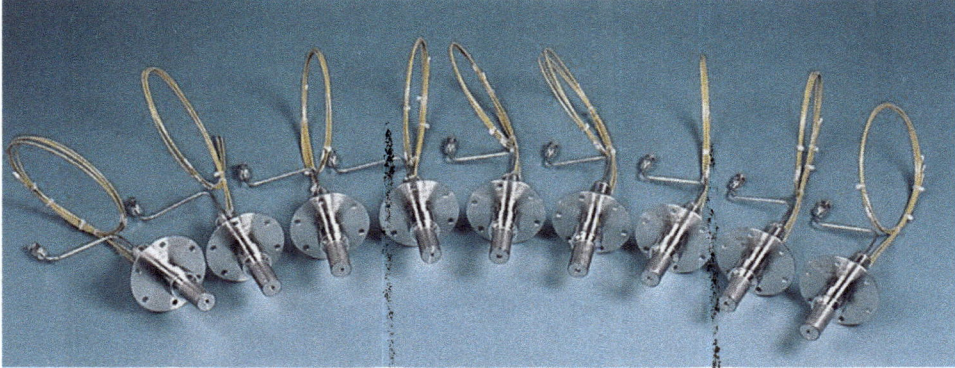

Figure 5.3. The hollow cathode assembly in the PCU on the ISS. (Courtesy of NASA.)

replacement board to the space environment and to possible triboelectric charge build-up on astronauts' suits and gloves (due to possible contact and separation of suit and glove surfaces). All four Hubble Servicing Missions required the astronauts to use the Shuttle robotic arm to position the telescope in the Shuttle's payload bay. Thus this repair was going to be performed in the wake, where the plasma density is not high enough to neutralize triboelectrically generated charges on those insulating surfaces. Extensive high-vacuum testing at NASA showed that triboelectric charging at levels that could generate discharges damaging to the boards were possible [9]. As a result, a special tool was developed to handle the boards. There were no discharge incidents during the mission.

5.5 Spacecraft charging in MEO

Although most satellites are placed in LEO or GEO, there are currently about 70 satellites orbiting in MEO [10]. Since the Van Allen radiation belts occupy the region between 2000 km and 19 000 km, most MEO satellites are GPS satellites orbiting just beyond the belts, about 20 000 km up. However, several scientific satellites, such as the NASA THEMIS mission satellites, have highly elliptical orbits through the MEO region and beyond (figure 5.4). High-speed electrons in the outer radiation belt are a hazard for these satellites as well as for any spacecraft crossing the radiation belts on the way to Mars or to other Solar System destinations.

As the solar wind stretches the long tail of the magnetosphere past the orbit of the Moon, it becomes overly stretched, snapping back, accelerating plasma electrons with it. These fast moving electrons emit electromagnetic waves called *chorus* waves, because they sound like singing birds when the signal is played through loudspeakers. These chorus waves accelerate electrons in the outer radiation belt to near the speed of light. Two of the THEMIS satellites (*D* and *E*) discovered that when the chorus waves interact with the plasma in the radiation belt, they lose energy and turn into *hiss* waves, so named because they make that sound when played through audio equipment. Hiss waves deflect high-energy electrons in the upper radiation belt toward the Earth, where they lose energy as they interact with atoms in the atmosphere [11].

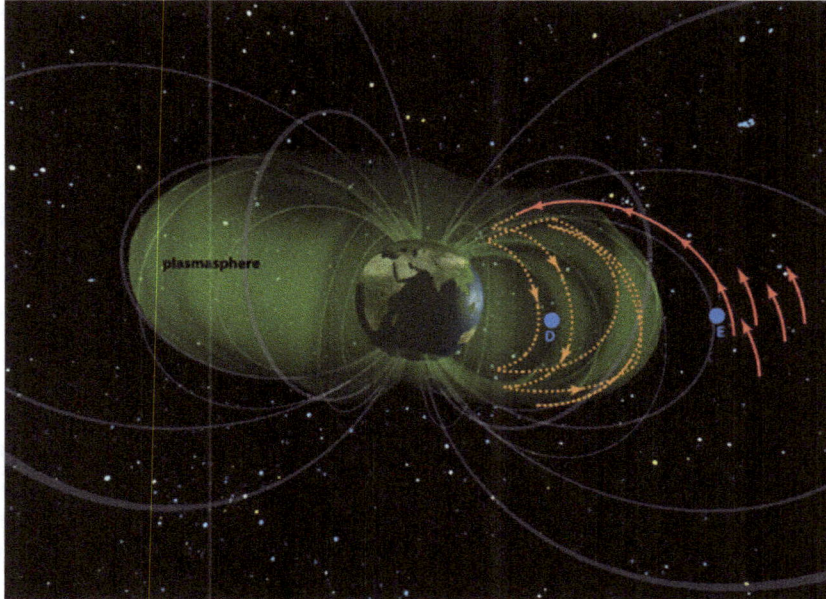

Figure 5.4. Elliptical orbits of THEMIS *D* and *E*, two of the five NASA satellites mapping the magnetosphere. These satellites discovered that chorus waves (red lines) generated by the magentotail change into hiss waves (yellow lines) that kill energetic electrons from the outer radiation belt. (Courtesy of NASA.)

In MEO, an issue that is probably more important than spacecraft charging is *deep dielectric charging*. Electrons and ions with sufficient energies can penetrate spacecraft surfaces, causing internal disruptions in electronics systems. Electrons with energies of a few kilo-electronvolts can penetrate about 1 μm into a dielectric [12]. Electrons with energies above 50 keV can penetrate spacecraft metallic surfaces [13]. In the radiation belts, electron energy can reach the mega-electronvolt range. Figure 5.5 indicates the mean penetration energies of protons and electrons into aluminum [1].

5.6 Spacecraft charging in GEO

GEO is the most important region for spacecraft charging because the plasma energy can be high and because of the high density of satellites in this region (about 40% of all satellites are in GEO). GEO also resides in the magnetosphere but outside the radiation belts. Satellites in GEO are close to the *magnetopause*, the boundary between the Earth's magnetic field and the solar wind. During strong solar activity, the front of the magnetosphere is pushed in past the geosynchronous orbits, a phenomenon that takes place several times a year.

On the night side of the Earth, at opposite ends to the magnetosphere's nose, spacecraft are immersed in the night side plasma. Due to its low-density (about 1 elementary charge cm^{-3}), this high-energy plasma is not able to neutralize any charge that builds up on spacecraft surfaces. Spacecraft surface in darkness can reach potentials between −10 kV and −30 kV. On the lit side, photoelectron

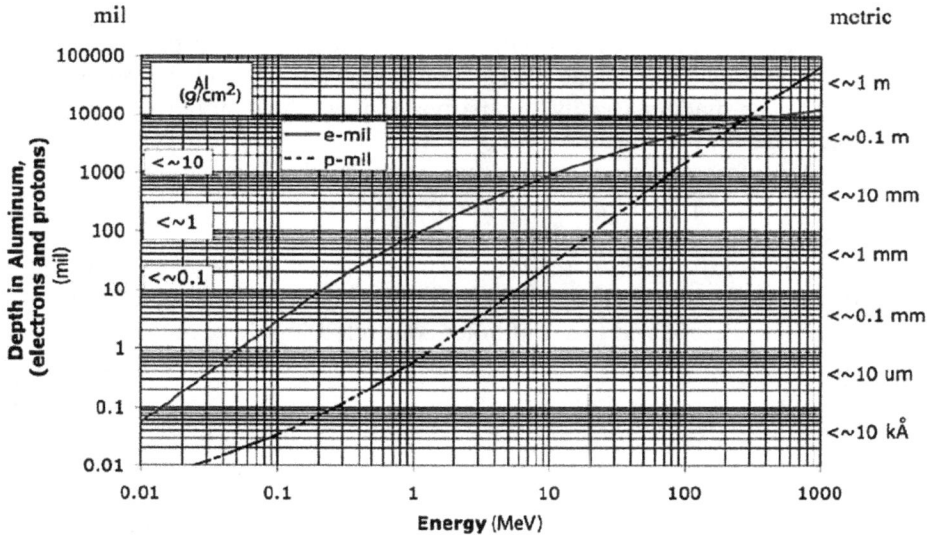

Figure 5.5. Penetration depth into an aluminum slab as a function of energy for electrons (solid line) and protons (dotted line) [1]. (Courtesy of NASA.)

emission due to the incident solar UV radiation results in surface potentials of +2 V to +5 V.

The large potential difference between the lit and dark sides of a spacecraft can cause differential charging. If the skin of the spacecraft is not electrically conductive, a potential difference of about 2 kV to about 11 kV builds up between the two sides. For aluminum or other electrically conductive surfaces, the potential difference is lower, of the order of 1 kV or less.

Of greater concern in this region is deep dielectric charging. Electrons with energies above 10 keV can penetrate insulating materials or pass through insulating films. A large charge buildup in an insulator can result in a rapid discharge that may damage onboard equipment.

5.7 Mitigation techniques

Space agencies around the world as well as spacecraft designers have developed several techniques to mitigate spacecraft charging. Spacecraft can have metallic surfaces, usually made of aluminum, as well as electrically insulating surfaces. The *NASA Technical Handbook* 'Mitigating in-space charging effects' [1] has charge mitigation recommendations for both metallic and non-metallic spacecraft surfaces.

For metallic surfaces, [1] recommends bonding (grounding) all metallic structural elements to the chassis with a leakage resistance of less than 10^8 ohms. In addition, all structural and mechanical parts should be electrically bonded to each other with a DC current resistance of less than 2.5×10^{-3} ohms at each joint.

For non-conducting or dielectric surfaces, [1] recommends conducting extensive ground analyses of the dissipation of charge to ground for all non-conductors to be used, to show that they will not present a hazard to the spacecraft. The *GEO*

Spacecraft Charging Guidelines [14] recommends coating all non-conductive surface materials on a spacecraft with a conductive material that can be grounded.

Dielectric materials used inside the spacecraft, such as wire insulation and circuit boards, should be made statically dissipative; that is, they should be able to leak charge faster than they can acquire it. These statically dissipative materials should be bonded to the spacecraft structure. Indium tin oxide (ITO) is a common statically dissipative coating that can be used on many dielectrics. ITO has the advantage of being transparent (it is commonly used in touch screens).

For large spacecraft—such as the ISS—plasma contactors are an effective method to control charging, as mentioned earlier. Simpler systems to ground large spacecraft, using micrometer-sized tips and holes connected to a bias potential, that emit electrons without the need for a gas have been developed but are not yet in place.

References

[1] NASA 2011 Mitigating in-space charging effects—a guideline *NASA Technical Handbook* NASA-HDBK-4002A

[2] Ferguson D C and Hillard G B 2003 Low Earth orbit spacecraft charging design and guidelines *NASA Technical Report 2003–212287*

[3] Anderson P C 2012 Characteristics of spacecraft charging in low earth orbit *J. Geophys. Res.* **117** A07308

[4] James B F *et al* 1994 The natural space environment: effects on spacecraft *NASA Reference Publication* 1350

[5] Tang D *et al* 2015 Particle-in-cell simulation study on the floating potential of spacecraft in the low earth orbit *Plasma Sci. and Technol.* **17** 288–93

[6] Koontz S *et al* 2003 Assessment and control of spacecraft charging risks on the international space station *8th Spacecraft Charging Technology Conf.* (*NASA Marshall Space Flight Center, Huntsville, AL, 20–23 October 2003*)

[7] Koontz S *et al* 2007 Progress in spacecraft environment interactions: International Space Station (ISS) development and operations *International Space Development Conf.* (*Dallas, TX, 25–28 May 2007*)

[8] Carpenter C B 2004 On the operational status of the ISS plasma contactor hollow cathodes *NASA Technical Report NASA/CR-2004-213184*

[9] Buhler C R, Clements J S and Calle C I 2012 Electrostatics studies for the 2008 Hubble Repair Mission *Proc. 2012 Joint Electrostatics Conf.* (*Cambridge, ON*)

[10] Horne R B 2015 Space weather charging environments especially radiation belts *SEREN Workshop on Physical Pathways to Space Weather Impacts* (*London*)

[11] Li W *et al* 2010 THEMIS analysis of observed equatorial electron distributions responsible for the chorus excitation *J. Geophys. Res.* **115** A00F11

[12] Tascione T E 1988 *Introduction to the Space Environment* (Malabar, FL: Orbit)

[13] Robinson P A and Coakley P 1992 Spacecraft charging: progress study of dielectrics and plasmas *IEEE Trans. Electr. Insul.* **27** 944–60

[14] Purvis C K, Garrett H B, Whittlesey A C and Stevens N J 1984 Design guidelines for assessing and controlling spacecraft charging effects *NASA Technical paper* 2361

Chapter 6

The electrostatic environment of the Moon

The Moon has an extremely rarefied atmosphere, composed of small amounts of helium, argon, neon, ammonia, methane, and carbon dioxide at a pressure of 10^{-12} mbar. This value represents a good vacuum. It is about the same as the vacuum found where the ISS orbits. With a particle density of 10^6 molecules cm^{-3} near the surface (10^{13} times smaller than the particle density near the surface of the Earth), the lunar atmosphere has no influence on the electrostatic environment of the Moon. What partly determines this environment is the flux of charged particles from the solar wind and cosmic rays. The interaction of this flux with the regolith produces a complex electrostatic environment.

6.1 The lunar surface environment

The lunar surface is covered by a regolith composed of several meters of rock and granular material ranging in size from a few centimeters to a few nanometers in diameter. Meteoroid impacts have grounded the top layers of the regolith down to small particles. The average size of the Apollo dust samples is about 70 μm. The regolith also includes a significant fraction of dust in the micrometer and sub-micrometer scale [1]. However, the size distribution is not homogeneous over the surface of the Moon, as evidenced by Apollo images (figure 6.1). The variation in size distribution is due to the different lengths of exposure to space weathering. The Apollo 15 landing site, for example, was on the eastern edge of Mare Imbrium, in the fine dust of the impact ejecta. The Apollo 16 landing site was in the lunar highlands, where rocks rich in the mineral feldspar are found.

6.2 The lunar electrostatic environment

Since the Moon does not have a substantial atmosphere, the lunar surface is directly exposed to the solar wind, to cosmic rays, and to solar UV radiation. The solar wind plasma is a flow of electrons and ions with a density of about 5 elementary charges cm^{-3} moving at about 400 km s^{-1} [2]. The density of the

Figure 6.1. Differences in regolith size distribution is evident in these images of the Apollo 15 and Apollo 16 landing sites. (Courtesy of NASA.)

cosmic rays is about six orders of magnitude smaller. The formalism describing the complex lunar electrostatic environment that results from these fluxes was partially developed by Manka [3] and completed by Halekas *et al* [4] and by Farrell *et al* [5]. The basic premise of this formalism is simply the law of conservation of electric charge: in equilibrium, the lunar surface will develop a charge so that the sum of all current fluxes cancels out. These current fluxes are: electrons and protons in the solar wind; protons (~90%) alpha particles (~9%) electrons (~1%), and a small fraction of heavy nuclei in cosmic rays; photoelectrons [6]; reflected and back-scattered electrons; secondary electrons from the lunar surface; solar wind ions and neutral particles reflected from the lunar surface [7–9]; neutrons generated in nuclear reactions with energetic incident radiation, and galactic cosmic rays (GCRs) partially reflected off the lunar surface [10]. Electrostatic potential contributions from these currents have been calculated using the Manka formalism for typical solar wind conditions using data from the Electron Reflectometer (ER) instrument on NASA's Lunar Prospector mission [4]. The ER instrument provided data on electron concentrations and temperatures. The plasma environment around Lunar Prospector's orbit was assumed to be neutral and the electrons and ions were assumed to have the same temperatures. These considerations showed that the majority of the contribution to lunar charging comes from only four currents: photoelectrons, solar wind electrons, solar wind ions, and secondary electrons [11].

The amount and sign of the charge that develops on the lunar surface is different for the sunlit and the night side of the Moon. On the sunlit side, the photoelectrons emitted by the under 200 nm UV radiation from the Sun as well as by solar x-rays leave the day side surface of the Moon positively charged. Photoelectrons with larger energies escape and join the plasma electrons surrounding the Moon. However, most photoelectrons have insufficient energy to escape, forming a sheath with a density of 10^3 to 10^4 electrons cm^{-3} about 1 m above the daylight lunar surface that shields the surface from the plasma [12]. A small positive potential of about 5–10 V would balance the photoelectron and incident solar wind electron currents [13]. Because of the shielding effect of the photoelectron sheath, the photoelectron current dominates on the sunlit side for the Moon.

On the night side, in the solar wind wake, the plasma density is substantially reduced, resulting in an increase in electron kinetic energy. Electrons in the solar wind plasma, moving faster than the ions, become the dominant current. This electron flux charges the lunar surface to potentials of about −50 to −200 V (of the order of the electron temperature) [14]. When the Moon crosses the Earth's plasma the potentials on the night side reach values of several hundred kilovolts. The Lunar Prospector spacecraft measured night side surface potentials of −4.5 kV during periods of intense solar activity [15]. A sheath extending to about 1 km from the surface is then formed on the lunar night side [12, 16].

A similar but smaller plasma wake is formed on the dark side of a crater in the lunar polar region where the solar wind flows horizontally over the crater. As on the night side, energetic plasma electrons move into this wake ahead of the ions, charging the surface negatively to hundreds of volts [17].

The lunar surface is then enveloped in a Debye sheath with a Debye length of the order of meters on the day side and of kilometers on the night side that shields the lunar surface potential (figure 6.2). Recall that the Debye length is the distance beyond which the electrostatic field of a charge is shielded by the opposite charges it has attracted.

The electrostatic fields near the *lunar terminator*, the boundary between the day and night sides, are very complex. In this region, the surface charge changes from

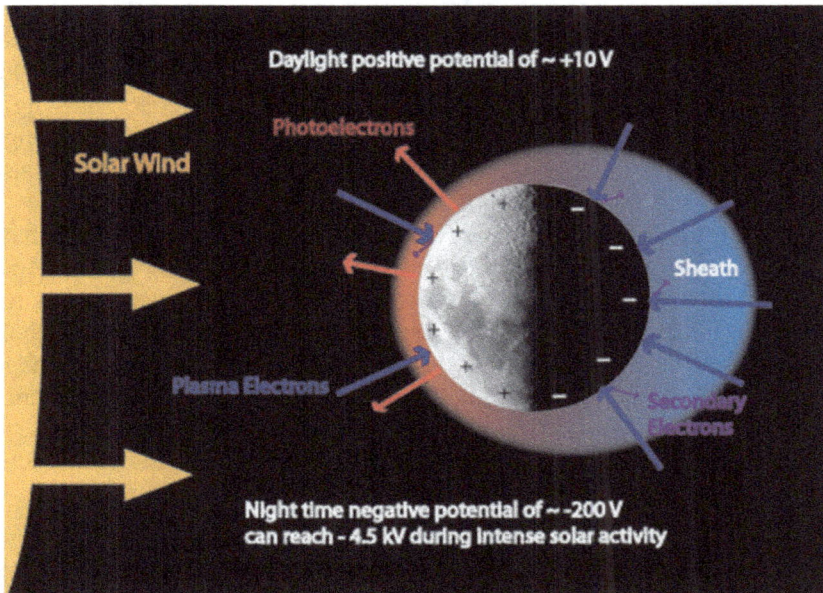

Figure 6.2. Lunar electrostatic environment.

positive in the day side to negative in the night side. Lunar Prospector measurements showed that the transition from the photoelectron sheath to the solar-wind plasma sheath takes place at a subsolar angle of about 67°. At this location, the outward photoelectron emission current balances the inward solar wind electron current and the potential changes from positive to negative. At the terminator (90° subsolar angle), the potential is about −40 V [5, 11].

Although fairly small compared to the photoelectron and plasma electron currents, secondary emissions should still be considered. Secondary radiation is due to interactions between incident GCRs and the atoms in the lunar regolith. The Cosmic Ray Telescope for the Effects of Radiation (CRaTER), an instrument on NASA's Lunar Reconnaissance Orbiter (LRO), measured space radiation at the Moon to determine protection levels for human missions. These measurements were performed during the longest solar minimum ever recorded. Many of the energetic electrons and nuclei in GRCs are deflected by the Sun's magnetic field. However, during a solar minimum, the weakened solar wind does not carry the solar magnetic field as far, allowing for a larger flux of GRCs which, in turn, generate a flux of secondary radiation about 30–40% higher than expected [18]. This amount is relatively small and is comparable to the annual occupational exposure for x-ray technicians.

6.3 Electrostatic charging of the lunar regolith

GCR particle radiation incident on the lunar surface can generate deep dielectric charging of the regolith down to a depth of about 1 m [18]. Since the lunar regolith has a very low electrical conductivity, electrostatic discharges can take place through the affected section of the regolith. NASA's Acceleration, Reconnection, Turbulence and Electrodynamics of the Moon's Interaction with the Sun (ARTEMIS) mission (figure 6.3), launched in 2010, is measuring the solar radiation incident on the lunar surface as the Moon moves in and out the Earth's magnetic field. Reka Winslow and collaborators are applying ARTEMIS data to a deep dielectric charging model developed by Jordan et al [19] to estimate the subsurface electric field strength and the possibility of dielectric breakdown in the lunar regolith [20]. GCRs can also alter the chemical composition of the regolith [19].

The complex lunar electrostatic environment affects surface dust behavior in unexpected ways. There has been some evidence of a horizon glow on the Moon that has been interpreted to be due to dust levitation and dust transport. This phenomenon was first observed during the NASA Surveyor missions to the Moon in the 1960s. The Surveyor 5, 6, and 7 lunar landers obtained television images of a glow just above the lunar horizon [21]. Because of its relatively high concentration, this lunar horizon glow could not be attributed to meteorite impacts. It was proposed instead that this glow was due to scattered sunlight from a cloud of submicron dust particles levitated by electrostatic forces up to a height of about 1 m above the surface near the terminator [22].

Additional evidence of this horizon glow and of dust levitation came during the Apollo missions. Apollo 17 astronauts sketched observations from orbit of what they referred to as 'streamers', 'bands', or 'twilight rays' on the lunar horizon

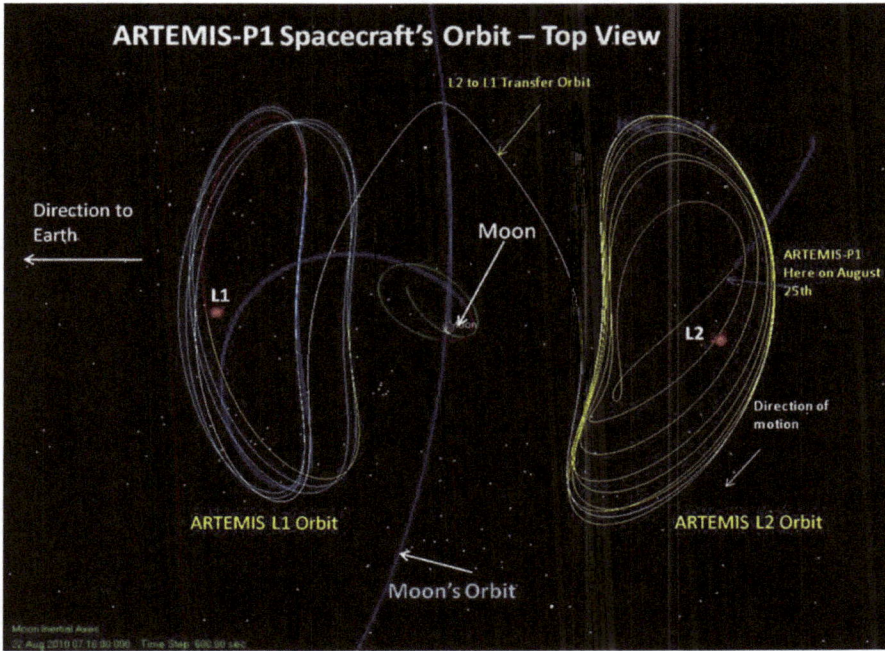

Figure 6.3. The ARTEMIS spacecraft's orbit. (Courtesy of NASA.)

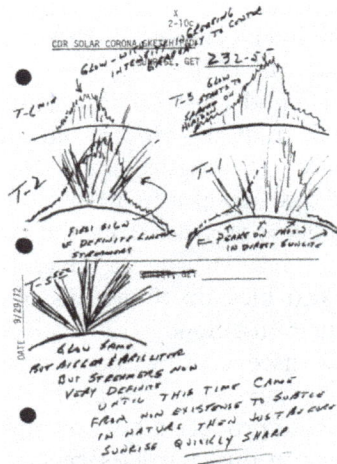

Figure 6.4. Sketches drawn in lunar orbit by Apollo 17 astronauts showing streamers reaching up to about 100 km from the surface. (Courtesy of NASA.)

(figure 6.4) [21]. A more recent analysis of these sketches puts the radius of the particles at less than 10 μm [24]. The Lunar Ejecta and Meteorites (LEAM) experiment on Apollo 17 detected the presence of dust clouds as well as evidence of slow-moving, highly charged dust particles [25].

Figure 6.5. Faint glow above the brighter glow caused by interplanetary dust particles observed during the Clementine mission. A definite interpretation of this faint glow has not been achieved. (Courtesy of NASA.)

The 1994 Clementine mission contained a dedicated Lunar Horizon Glow experiment that required the spacecraft to be in the lunar umbra. Unfortunately, due to the geometry of the trajectory relative to the Sun, conflicts with the more important mapping experiments onboard and data downlinks prevented the Lunar Horizon Glow experiment from being carried out [26]. However, images with the spacecraft's tracker camera detected a faint glow above the brighter glow of zodiacal dust particles (figure 6.5). The interpretation of these images is complicated and has never been completed satisfactorily [27].

NASA's Lunar Atmosphere and Dust Environment Explorer (LADEE), which launched in September 2014 to study the Moon from orbit until it was intentionally crashed on the lunar surface six months later, discovered that the Moon is surrounded by a permanent dust cloud lofted from the surface by the impacts of interplanetary dust particles. These particles strike the surface at speeds of 34 km s^{-1}, vaporizing part of the soil and releasing a large amount of heat. A single interplanetary dust particle can kick up thousands of surface dust particles. The total estimated mass of the lunar dust cloud is 120 kg. However, LADEE did not find evidence for the horizon glow discovered by Surveyor spacecraft and the Apollo 17 mission. Submicron particle densities were measured at less than 100 particles m^{-3}, a value considered to be too low to produce a horizon glow [28]. Thus the issue of the electrostatic lofting of submicron dust particles generating a horizon glow on the Moon remains unresolved.

6.4 Triboelectric charging on the lunar surface

An object moving on the surface of the Moon will charge due to contact with the surface as well as to the exposure to the surrounding plasma environment. Given the possible high potentials on the lunar night side during periods of intense solar activity, this triboelectric charging may affect human missions to the lunar surface.

A recent study by Jackson *et al* [29] indicates that the time it takes for charge to dissipate depends on the local plasma environment. They performed modeling analyses with the rubber tires on the Modular Equipment Transporter (MET) used during the Apollo 14 mission and with the aluminum and titanium tires on the Lunar Roving Vehicle (LRV) used during Apollo 15, 16, and 17.

On the lunar day side, a rover is exposed to the same electrostatic environment as the lunar surface, emitting photoelectrons with energies that depend on the work function of the material. Calculations for the MET insulating wheel under lunar daylight conditions showed that it would charge to a potential similar to the ~3 V potential of the lunar surface. The metallic wheel on the LRV would charge to 4.62 V. These low potentials would decay very quickly (in less than 0.5 ms) to reach equilibrium with the surrounding environment [29]. As the rover moves along the lunar surface, repeated wheel–regolith contact and separation would generate tribo-electric charging on both surfaces. Results from modeling the motion of the two wheels using a charge continuity equation developed by Farrell *et al* [30] for several speeds ranging from 0.01 m s^{-1} to 3 m s^{-1} show that in the photoelectron dominated daylight plasma region, the charge dissipates almost instantly in all cases. The photoelectron plasma region is conductive and triboelectric charging on the lunar day side is not an issue.

On the lunar night side, where surface potentials can regularly reach −200 V, a rover will be exposed to the more energetic solar wind electrons that dominate this region as well as to secondary electron currents. A similar situation occurs for a rover moving on the shadowed polar crater region, where the higher energy solar wind electrons and secondary electrons from the surface charge the surface negatively. At the higher electron speeds in these two regions, both types of rover wheels would charge to −1 kV, dissipating to equilibrium in about 10^7–10^8 s [29]. These large potentials and long dissipation times will pose a hazard to astronauts and equipment during lunar missions.

References

[1] Grün E *et al* 2011 The lunar dust environment *Planet. Space Sci.* **59** 1672–80

[2] Dessler A J 1967 Solar wind and interplanetary magnetic field *Rev. Geophys.* **5** 1

[3] Manka R H 1973 Plasma and potential at the lunar surface *Photon and Particle Interactions with Surfaces in Space* (New York: Springer) pp 347–61

[4] Halekas J S, Bale S D, Mitchell D L and Lin R P 2005 Electrons and magnetic fields in the lunar plasma wake *J. Geophys. Res.* **110** A07222

[5] Farrell W M, Stubbs T J, Vondrak R R, Delory G T and Halekas J S 2007 Complex electric fields near the lunar terminator: the near-surface wake and accelerated dust *Geophys Res. Lett.* **34** L14201

[6] Feuerbacher *et al* 1972 Photoemission from lunar surface fines and the lunar photoelectron sheath *Proc. Lunar Sci. Conf.* **3** 2655–63

[7] Saito Y S *et al* 2008 Solar wind proton reflection at the lunar surface: low energy ion measurements by MAPPACE onboard SELENE (KAGUYA) *Geophys. Res. Lett.* **34** L24205

[8] Weiser M *et al* 2010 First observation of mini-magnetosphere above a lunar magnetic anomaly using energetic neutral atoms *Geophys. Res. Lett.* **37** L01502

[9] McComas D J *et al* 2009 Lunar backscatter and neutralization of the solar wind: first observation of neutral atoms from the Moon *Geophys. Res. Lett.* **36** L12104

[10] Porter J A *et al* 2013 Radiation environment at the Moon: comparisons of transport code modeling and measurements from the CRaTER instrument *Space Weather* **12** 329–36

[11] Stubbs T J, Halekas J S, Farrell W M and Vondrak R D 2005 Lunar surface charging: a global perspective using lunar prospector data *Workshop on dust in planetary systems, Lunar and Planetary Institute* 4070

[12] Singer S F and Walker E H 1962 Photoelectric screening of bodies in interplanetary space *Icarus* **1** 7–12

[13] Halekas J S, Poppe A, Delory G T, Farrell W M and Horanyi M 2012 Solar wind electron interaction with the dayside lunar surface and crustal magnetic fields: evidence for precursor effects *Earth and Planets* **64** 73–82

[14] Halekas J S, Mitchell R P, Lin L L, Hood M, Acuña H and Binder A 2002 Evidence for negative charging of the lunar surface in shadow *Geophys. Res. Lett.* **29** 1436

[15] Halekas J S, Delory G T, Stubbs J K and Farrell W M 2009 Lunar surface charging during solar energetic particle events *J. Geophys. Res.* **114** A05110

[16] Freeman J W and Ibrahim M 1975 Lunar electric fields surface potential and associated plasma sheaths *Moon* **14** 103–14

[17] Farrell W M *et al* 2008 Concerning the dissipation of electrically charged objects in the shadowed lunar polar regions *Geophys. Res. Lett.* **35** L19104

[18] Schwadron N A *et al* 2015 LRO/CRaTER discoveries of the lunar radiation environment and lunar regolith alteration by radiation *46th Lunar and Planetary Science Conf.* 2395

[19] Jordan A P, Stubbs T J, Wilson J K, Schwadron N A and Spence H E 2015 *J. Geophys. Res.* **120** 210–25

[20] Winslow R M *et al* 2015 Lunar surface charging and possible dielectric breakdown in the regolith during two strong SEP events *46th Lunar and Planetary Science Conference* 1261

[21] Criswell D R 1973 Horizon glow and the motion of lunar dust *Photon and Particle Interactions in Space* ed R J L Gerard (Dordrecht: D Reidel) p 545

[22] Rennilson J J and Criswell D R 1974 Surveyor observations of lunar horizon glow *The moon* **10** 2

[23] Zook H A and McCoy J E 1991 Large scale lunar horizon glow and a high altitude lunar dust exosphere *Geophys. Res. Lett.* **18** 2117

[24] Stubbs T J, Vondrak R R and Farrell W M 2006 A dynamic fountain model for lunar dust *Adv. Space Res.* **37** 59–66

[25] Berg O E, Wolf H and Rhee J 1975 Lunar soil movement registered by the apollo 17 cosmic dust experiment *Interplanetary and Zodiacal Light* 233 (Heidelberg: Springer)

[26] Sorensen T C and Spudis P D 2005 The clementine mission: a 10-year perspective *J. Earth Syst. Sci.* **114** 645–68

[27] Hahn J M, Zook H A, Cooper B and Sunkara B 2002 Clementine observations of the zodiacal light and the dust content of the inner solar system *Icarus* **158** 360–78

[28] Szalay J R and Horanyi M 2015 The search for electrostatically lofted grains above the moon with the lunar dust experiment *Geophys. Res. Lett.* **42** 5141–6

[29] Jackson T L, Farrell W M and Zimmerman M I 2015 Rover wheel charging on the lunar surface *Adv. Space Res.* **55** 1710–20

[30] Farrell W M *et al* 2010 Anticipated electrical environment within permanently shadowed lunar craters *J. Geophys. Res.* **115** E03004

Chapter 7

The electrostatic environment of asteroids

Asteroids are small, irregular, airless, rocky Solar System bodies ranging in size from meters to about 1000 km. Most asteroids orbit the Sun in the *Asteroid Belt*, a region between the orbits or Mars and Jupiter, extending from 2.2 AU to 3.3 AU (figure 7.1). Their periods of orbital revolution around the Sun range from 3.3 to 6 years. There are tens of thousands of these bodies but their total mass, at 8×10^{20} kg, is less than that of the Moon. The typical distance between asteroids larger than 1 km is of the order of several million kilometers.

There are three classes of asteroids: C-type asteroids (chondrite) are made up of clay and silicate rocks, are dark, and are among the oldest bodies in the Solar System; S-type (stony) are made up of silicone materials and nickel–iron; and M-type (metallic) are made up of nickel–iron. More than eighty percent of the asteroids found at the outer edges of the asteroid belt are C-type asteroids. Their composition is similar to that of the carbonaceous chondrite meteorites that fall to Earth. In fact, these meteorites are probably pieces formed during asteroid collisions. S-type asteroids are found mostly at the inner edge of the asteroid belt and are the most abundant [1]. The M-type asteroids are the least common and are found in the middle region of the asteroid belt.

Flybys of asteroids Gaspra and Ida by NASA's Galileo mission as well as spectral measurements of S-type asteroids, the most common type, indicate that lunar-like space weathering, such as solar wind sputtering and micrometeorite bombardment, occurs on asteroids [1]. However, in the asteroid belt these processes are different. The solar wind flux is reduced at the larger distance from the Sun and micrometeorite impacts are slower but more frequent [2]. Images taken by the NASA Near Earth Asteroid Rendezvous (NEAR)-Shoemaker spacecraft of the asteroid 433 Eros, one of the largest near-Earth asteroids, show evidence of a layer of fine dust, with grain sizes under 50 μm in diameter (figure 7.2).

Figure 7.1. Image of the asteroid Eros, an S-type asteroid, taken by NASA's NEAR mission in 2000. (Courtesy of NASA.)

Figure 7.2. The dusty surface of Eros as photographed by NEAR-Shoemaker from a height of 250 m. The image is 12 m across. (Courtesy of NASA.)

7.1 The asteroid electrostatic environment

Like the Moon, asteroids lack an atmosphere and are therefore fully exposed to the solar wind, solar UV radiation, and cosmic rays. Arguments similar to those for the Moon indicate that the interaction of the sunlit and night sides of an asteroid with these charging currents would result in a Debye sheet that surrounds and shields the asteroid surface. On the sunlit side, photoelectron emission currents prevail, while on the night side, solar wind electron currents are dominant. Following the approach of Manka [3] and of Mendis *et al* [4], Lee calculated the equilibrium surface potential at the sunlit surface of a main belt asteroid to be about +5 V [5]. On the night side surface, calculations show that the potential is about −200 V but can reach -10^3 V [2, 6]. Along the terminator, where the photoelectron emission and solar wind electron currents balance, surface potentials change gradually from positive to negative.

7.2 Electrostatic dust transport

The dusty asteroid regolith may become charged by the different currents reaching the surface. The different electrostatic fields present near the asteroid surface may transport these charged particles in different ways. In 1996, Lee predicted that electrostatic forces could levitate and transport dust on asteroids to 'smooth, flat, and/or permanently shaded areas' [5]. In 2001, images from the NEAR-Shoemaker spacecraft showed smooth areas in craters 20–300 m in diameter that were interpreted as 'ponds' [7]. The smoothness of the ponds as well as the colors of the pond material suggest that they are covered with dust grains 50 μm and under in diameter, consistent with Lee's electrostatic model [8].

Dust particles on the sunlit side of an asteroid may have a positive charge due to photoelectron emission. The electrostatic field in the photoelectron sheet, which points away from the asteroid surface, may levitate some of these positively charged particles. Depending on the particle's mass, the magnitude of its positive charge, its initial velocity, and the asteroid's gravity, the particle may stay inside the sheath or cross the layer. If the particle stays in the sheath, it may acquire a negative charge due to photoelectron absorption, which is greater than photoelectron emission. If the particle crosses the photoelectron sheath, it will emit photoelectrons and become positively charged [9]. It may then continue moving outward and leave the asteroid, fall back into the sheath, or even reach an equilibrium state, where the electrostatic force balances its weight.

Across the terminator or across boundaries between sunlit and shadow regions in craters, horizontal electric fields can form. In these cases, the electric fields will emanate from the positively charged sunlit side to the negatively charged shadow region. Charged dust can then be transported across these regions. A study of the particle dynamics on the asteroid Eros by Colwell *et al* showed that electrostatic levitation of dust may provide enough dust to roughly account for the formation of the ponds observed by the NEAR-Shoemaker spacecraft [9].

7.3 Cohesive forces in asteroids

Analyses of the distribution of asteroid spin rates versus size have shown that asteroids larger than a few hundred meters in diameter are in general loosely bound aggregates with minimal tensile strength (so-called *rubble pile* asteroids) [10]. Smaller asteroids, with diameters under 100 m, are monolithic. The JAXA Hayabusa mission to asteroid Itokawa, which has a mean radius of 162 km, successfully landed a spacecraft on the asteroid in November 2005 and collected samples. Measurements of those samples and estimates of the sizes of rocks showed that the size distribution of the regolith components ranges from under 100 μm to millimeters in diameter for granular material and from millimeters to 40 m for rocks and boulders [11]. Observations of Main Belt asteroids have shown that regolith dust sizes extend down to about 10 μm [12].

Recent studies have looked at the nature of the cohesive forces keeping the aggregate material of an asteroid together. Measurements of the rotation period and bulk density of the near-Earth asteroid 1950 DA indicate that it is rotating faster than gravity and friction would allow [13]. This asteroid has a mean diameter of 1.3 km and a rotation period of 2.1216 h, a value higher than the estimated critical spin limit of 2.2 h for a coherent or monolith asteroid [14]. The measured rotation period clearly requires the presence of cohesive forces. The possible forces are electrostatic and van der Waals forces between grains. The van der Waals force between two neutral objects is due to the interaction between induced dipoles that result from polarizable atoms in the objects. The van der Waals force decreases rapidly with separation between objects, becoming negligible at distances over 30 nm or so. When electrostatic forces are present, as is usually the case, these two interactions have about the same strength at this distance of 30 nm. At greater separations, the electrostatic force prevails. At 10 nm, van der Waals has about ten times the strength of the electrostatic force [15]. Van der Waals forces are significant only for particles with diameters under 100 μm and predominate for particles under 10 μm [16]. Thus, besides gravity and friction, the cohesion forces holding together rubber pile asteroids are probably van der Waals forces between fine grains that may act as the glue that holds rocks and boulders together along with electrostatic forces acting on the larger grains and rocks.

For the 1950 DA asteroid, Rozitis *et al* found that a minimum cohesive strength of 64 Pa is required to keep the asteroid intact, which is in the 3–300 Pa range determined by numerical simulations and in rough agreement with the value of 100 Pa found for lunar regolith [13, 17–18]. Sanchez and Scheeres ran a simulation of a boulder immersed in regolith using a 1 m boulder half-immersed in a hemispherical container filled with particles between 2 and 3 cm in diameter [17]. Using a value of 1 milli-g for the gravitational field and actual asteroid spin data, they obtain a value of 25 kPa for the cohesive strength of a rubble pile asteroid. This value is consistent with rubble piles containing dust particles with diameters under 10 μm acting as glue for larger boulders.

References

[1] Noble S K *et al* 2010 Evidence of space weathering in regolith breccias II: asteroidal regoligh breccias *Meteorit. Planet. Sci.* **45** 2007–15

[2] Hortz F and Schall R B 1981 Asteroidal agglutinate formation and implications for asteroidal surfaces *Icarus* **46** 337–53

[3] Manka R H 1973 Plasma potential at the lunar surface *Photon and Particle Interactions with Surfaces in Space* ed R L Grad (Dordrecht: Reidel) pp 347–61

[4] Mendis D A *et al* 1981 On the electrostatic charging of the cometary nucleus *Astrophys. J.* **249** 787–97

[5] Lee P 1996 Dust levitation on asteroids *Icarus* **124** 181–94

[6] Halekas J S, Mitchell D L, Lin R P, Hood L L, Acuña M H and Binder A 2002 Evidence for negative charging of the lunar surface in shadow *Geophys Res. Lett.* **29** 1436

[7] Veverka J *et al* 2001 Imaging of small-scale features on 433 Eros from NEAR: evidence for a complex regolith *Science* **292** 484–8

[8] Robinson M S *et al* 2001 The nature of ponded deposits on Eros *Nature* **413** 396–400

[9] Colwell J E *et al* 2005 Dust transport in photoelectron layers and the formation of dust ponds on Eros *Icarus* **175** 159–69

[10] Pravec P and Harris A W 2000 Fast and slow rotation of asteroids *Icarus* **148** 12–20

[11] Tsuchiyama A *et al* 2011 Three-dimensional structrure of Hayabusa samples: origin and evolution of Itokawa regolith *Science* **333** 1125–8

[12] Jewitt D *et al* 2010 A recent disruption of the Main Belt asteroid p/2010 a2 *Nature* **467** 817–9

[13] Rozitis B, MacLennan E and Emery J P 2014 Cohesive forces prevent the rotational breakup of rubble-pile asteroid (29075) 1950 DA *Nature* **512** 174–6

[14] Harris A W 1996 The rotation rates of very small asteroids: evidence for "rubble pile" structure *Proc. Lunar Planet. Sci. Conf. 27th* 493–4

[15] Gady B *et al* 1996 Identification of electrostatic and van der Waals interaction forces between a micrometersize sphere and a flat surface *Phys. Rev.* B **53** 8065–70

[16] Casterllanos A 2005 The relationship between attractive interparticle forces and bulk behaviour in dry and uncharged fine powders *Adv. Phys.* **54** 263–376

[17] Sanchez P and Scheeres D J 2014 The strength of regolith and rubble pile asteroids *Metorit. Planet. Sci.* **49** 788–811

[18] Mitchell J K *et al* 1974 Apollo soil mechanics experiment S-200 final report *Space Sci. Lab. Ser.* **15** 72–85

Chapter 8

The Martian electrostatic environment

8.1 The Martian atmosphere

The Martian atmosphere is composed of 95.32% carbon dioxide, 2.7% nitrogen, 1.6% argon, 0.13% oxygen, 0.08% carbon monoxide, and traces of water (0.02%), nitrogen oxide, neon, krypton, and xenon. The average atmospheric pressure near the surface of the planet is 6.36 mb (4.77 Torr), ranging from 4.0 to 8.7 mb, which is about 0.6% of the Earth's atmospheric pressure near the surface. The density of the Martian atmosphere decreases by a factor of $1/e$ relative to the density at the surface at an altitude of 11.1 km. The altitude at which the density and pressure of an atmosphere decreases to this factor is known as the *scale height* of the atmosphere. The scale height of the Earth's atmosphere is 8.6 km.

Like the Earth, the atmosphere of Mars is divided into distinct layers (figure 8.1). Closest to the surface is the troposphere, where convection takes place. The height of the troposphere changes with surface temperature. During the day in the Martian summer, surface temperatures can reach 300 K and the height of troposphere can reach 30 km. During night time in the polar regions, where the temperature is lowest, the troposphere disappears altogether. Above the troposphere is the stratosphere, where the temperature stays fairly constant up to about 140 km. At higher altitudes, the temperature rises steadily.

Water ice clouds occasionally form in the troposphere, usually around mountains. Higher up, in the stratosphere, carbon dioxide condenses and forms clouds and haze.

Dust is uploaded into the atmosphere from the planet's dusty surface by a mechanism that is not completely understood. The layer of dust and sand that covers the planet has been somewhat homogenized by global dust storms that take place every few Martian years. Dust devils up to 10 km high are relatively common on Mars. Some of the landing locations for the Mars Exploration Rovers (MERs) have detected dust devils in their vicinity almost daily. It is perhaps this dust storm activity which generates the mechanism that raises dust into the tenuous Martian

8-1

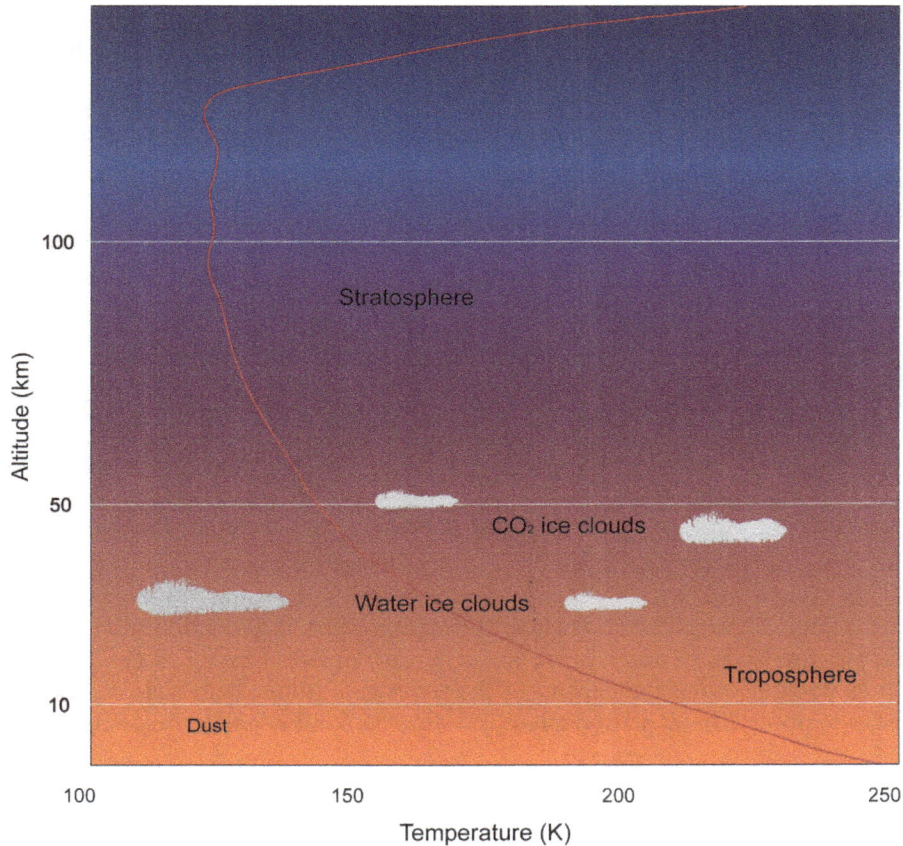

Figure 8.1. Temperature variation with altitude in the Martian atmosphere.

atmosphere. It has been proposed that saltation in the presence of electric fields may uplift the dust [1, 2].

Landis *et al*, analyzing data from the Microscopic Imager on the NASA MERs, have found that the dust content of the atmosphere has a three-component particle distribution: *atmospheric dust* suspended for long periods of time, with diameters from 2 to 4 μm; *settled dust* uplifted into the atmosphere by dust devils and wind storms, with particle diameters of 10 μm and under; and *saltating particles*, with diameters over 80 μm [3]. The Microscopic Imager also measured the average diameter of particles on the surface of Mars to be 220 μm.

Although the density of dust in the Martian atmosphere has never been measured directly, these values can be obtained from measurements of the opacity of the atmosphere that have been taken from landers. The opacity of the atmosphere is measured in terms of the *optical depth* τ, which is a measure of the transmission of radiation through the atmosphere. Optical depth is given by the logarithm of the ratio of transmitted to incident radiant power through the atmosphere. Typical values of the optical depths in the Martian atmosphere during non-dust storm conditions range from 0.2 to 1 [4]. Optical depths during local dust storm conditions are in the range from 1 to 6. Figure 8.2 shows optical depths measured by the MERs Spirit and Opportunity during 5 years of their mission [4].

Using the MER optical depth data, we can calculate the expected atmospheric dust particle density for different conditions. The particle density as a function of height z can be approximated from [5]

$$N \sim N_0 \tau e^{-\frac{z}{H}}$$

where N_0 is the number density at the surface for an optical depth of 1 and H is the scale height, which has an average value of 11.1 km. For relatively clear atmospheric conditions, with the optical depth τ from 0.2 to 1, the average number of dust particles in the atmosphere near the ground $(z = 0)$ ranges from about 5–24 particles cm^{-3}. For dust storm conditions, using $\tau = 6$, the expected particle density is about 140 particles cm^{-3}.

By comparison, a typical terrestrial indoors environment is considered similar to a class 100 000 clean room, a classification used in the United States by the electronics, pharmaceutical, and medical industries [6]. A class 100 000 clean room contains 100 000 particles 0.5 μm and larger in diameter per cubic foot of air. This value is equal to 3.5 particles cm^{-3}, which falls at the low end of the range of the atmospheric particle density during non-dust storm conditions on Mars. However, the Martian atmosphere has a density of 0.020 kg m^{-3} near the surface, which is about 1.6% of the density of the terrestrial atmosphere near the surface. If we were to pump Martian atmospheric gas into a chamber and increase its density to

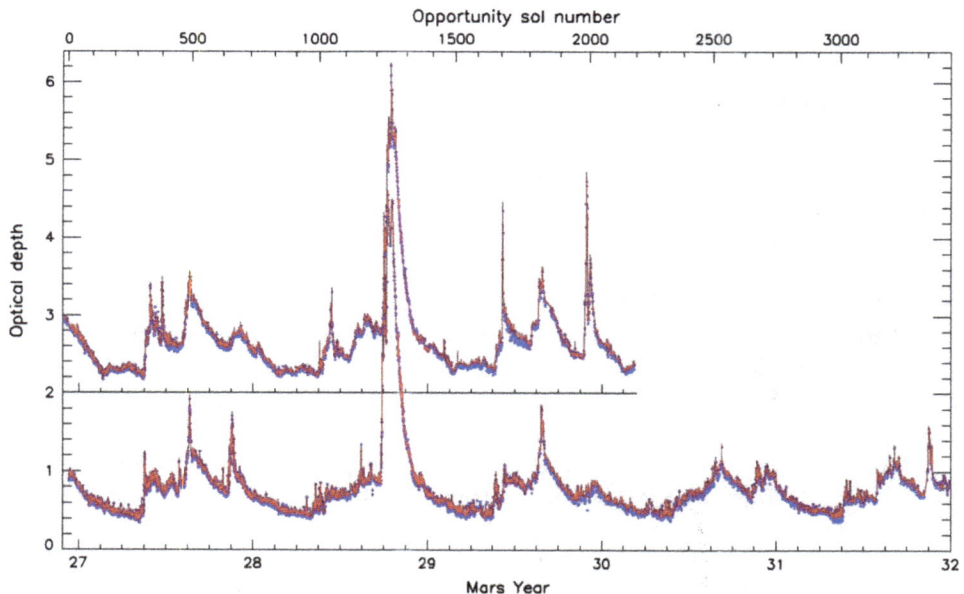

Figure 8.2. MERs' optical depth for the first 1200 sols. The opportunity data at the top is offset by 2 [4]. (Courtesy of NASA.)

match that of the Earth's atmosphere, the particle concentration would increase from an average of about 11 (taking the middle of the range for calm conditions) to about 670 particles cm^{-3}.

NASA's Mars human exploration program relies on the utilization of local resources for astronaut sustainability. A plant that would pump Martian atmospheric gas into a chemical reactor chamber for oxygen, methane, and water extraction would contain amounts of dust that could clog up filters and may damage reactors. Methods to use electrostatic forces to extract dust particles from these planned intakes are being studied [7].

8.2 Electrical breakdown in the Martian atmosphere

At about 0.6% of the Earth's atmospheric pressure, breakdown in the Martian atmosphere occurs at a much lower potential than on Earth. Since the Martian atmosphere is mostly made up of carbon dioxide, a quick glance at figure 8.3 indicates that with a 1 mm gap, breakdown in CO_2 will take place at about 450 V. Near the Earth's surface, breakdown with this gap requires 4.3 kV. Experimental data comparing breakdown in CO_2 and in a premixed gas that emulates the Martian atmosphere shows that the breakdown potentials are nearly identical (figure 8.3) [7]. These experiments were carried out in a vacuum chamber at 9 mb of pressure. This data indicates that the CO_2 dominates the Paschen breakdown and that the additional gases modify these potentials by an average of only 15 V.

8.3 Electrostatic charge and size of Martian atmospheric dust particles

Triboelectrification of dust grains during terrestrial dust storms or dust devils produces positively and negatively charged grains. On Mars, convective instabilities in the atmosphere should stratify similarly produced charged dust grains, with lighter grains being lifted to higher altitudes than more massive grains. Since smaller

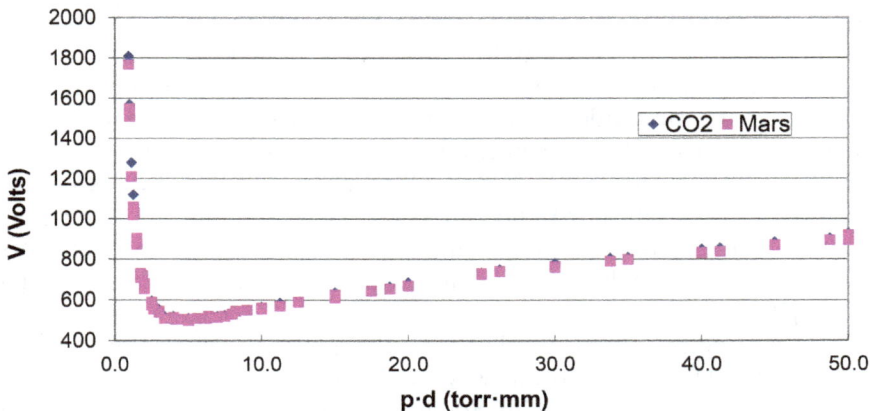

Figure 8.3. Paschen breakdown potentials versus pressure–distance for a Martian gas mixture (red squares) and for CO_2 (blue triangles) [7].

particles tend to charge negatively and larger particles charge positively [8], a macroscopic dipole moment is formed in the atmosphere that can produce an electrical discharge [9]. Fabian, Krauss, and their collaborators demonstrated experimentally in a simulated Martian atmosphere that this type of dust vertical motion can generate electric fields strong enough for electrical discharges to occur [10, 11].

Numerical models of dust electrification during Martian dust storms [12] and dust devils [13] predict that these electric fields should have strengths up to the theoretical breakdown potential of carbon dioxide at the low atmospheric pressure near the surface of Mars. Combined with experimental values of electron density in the Martian atmosphere, these models yield values of the electrical conductivity of the atmosphere that are several orders of magnitude higher than the values for the terrestrial atmosphere [14]. If these numbers are correct, charge dissipation in the Martian atmosphere would take place in seconds rather than minutes, as is the case for Earth. The discharge mechanism, however, remains unknown. Whether this discharge takes place violently, through lightning, or more gently, as in corona glow, is not known. No direct measurements of possible lightning or glow discharges in the Martian atmosphere have ever been made. However, there is some experimental evidence for glow discharge in laboratory experiments. Eden and Vonnegut placed sand particles in a container with carbon dioxide at pressures in the range of the Martian atmospheric pressure and observed a glow as well as filamentary electrical discharges when the container was shaken [15]. More recently, our NASA laboratory conducted similar experiments where we were able to observe a visible glow and show that these discharges altered several organics known to exist on Mars [16]. In contrast with these findings and earlier results, a more recent charging model showed that electric fields cannot reach levels up to breakdown because of charge dissipation in the saltation layer [17].

Searches for evidence of electrostatic discharges in the Martian atmosphere have been made with instrumentation aboard orbiting spacecraft. In 2009, Ruf and collaborators claimed that they had detected non-thermal electromagnetic emissions during a dust storm. Analyses of the modes of these emissions were interpreted to be Schumann resonances. Some researches attribute the presence of these resonances to lightning discharges. However, a subsequent observation in the same electromagnetic region found no evidence of Schumann resonances during a period that included dust storms [18]. Detailed studies of over 5 years of observations by the Mars Advanced Radar for Subsurface and Ionosphere Sounding (MARSIS) yielded no evidence of high-frequency radio emissions that would indicate the presence of electrical discharges. Moreover, the connection between Schumann resonances and lightning has not been established yet, with only one research effort indicating it as a possibility [19].

A key outstanding question related to the presence of lightning and glow discharges in the Martian atmosphere is the rate of charge dissipation in the more conductive Martian atmosphere [20]. Some terrestrial examples of particle charging in volcanic ash clouds have shown that they remain electrified long after charge should have dissipated into the atmosphere [21]. A similar phenomenon could

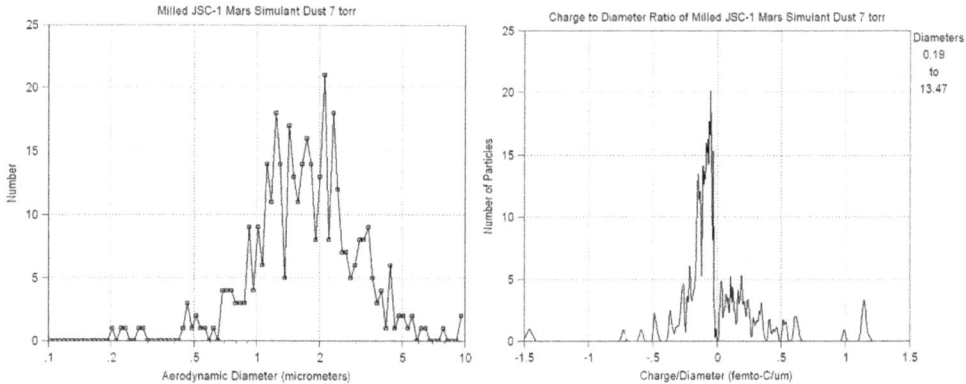

Figure 8.4. Aerodynamic diameter (left) and electrostatic charge density (right) of simulated Martian dust particles measured in a low-pressure carbon dioxide atmosphere. The diameters of the particles used are about half of the observed diameters on Mars [21].

happen on Mars, which may influence electrical activity. Ions and electrons present in the atmosphere may also be a factor in limiting the strength of the electric fields and the conductivity of the atmosphere [22].

Experiments to measure the charge and aerodynamic diameter of dust particles in a simulated Martian atmospheric environment were carried out in our NASA laboratory in collaboration with the University of Arkansas at Little Rock [23]. Electrostatic charge and aerodynamic diameters for simulated Martian dust particles were measured in a carbon dioxide atmosphere at 6 mbars of pressure (figure 8.4). The average particle diameter used in these experiments is smaller than the actual particle diameters derived from optical data taken from spacecraft in orbit around Mars and with lander instruments. The dust cross section weighted mean radius at Gusev crater where Spirit was operating was 1.47 ± 0.21 μm, and 1.52 ± 0.18 μm at Meridiani crater, Opportunity's home [24]. The experiments showed the feasibility of an instrument that could be deployed in a future mission to directly measure both particle diameter and electrostatic charge.

References

[1] Reno N O and Kok J F 2008 Electrical activity and dust lifting on Earth Mars and beyond *Space Sci. Rev.* **137** 19–134

[2] Greely R *et al* 2003 Martian dust devils *J. Geophys. Res.* **108** 5041

[3] Landis G A, Herkenhoff K, Greeley R, Thompon S and Whelley P 2006 and the MER Athena Science Team Dust and sand deposition on the MER solar arrays as viewed by the Microscopic Imager *Lunar Planet. Sci.* **37** 1937

[4] Lemmon MT *et al* 2014 Dust aerosol clouds and the atmospheric optical depth record over 5 Mars years of the Mars Exploration Rover mission *Icarus* **251** 96–111

[5] Haberle R M *et al* 1993 Atmospheric effects on the utility of solar power on Mars *Resources of Near Earth Space* (Tucson, AZ: University of Arizona Press)

[6] Health Facilities Institute 2011 Indoor air quality and particle measurement, http://www.healthyfacilitiesinstitute.com/a_128-Indoor_Air_Quality_and_Particle_Measurement

[7] Calle C I, Thompson S M, Cox N D, Johansen M R, Williams B S, Hogue M D and Clements J S 2011 Electrostatic precipitation of dust in the Martian atmosphere: implications for the utilization of resources during future manned exploration missions *J. Phys.: Conf. Ser.* **327** 012048

[8] Forward K M *et al* 2009 Particle-size dependent bipolar charging of Martian regolith simulant *Geophys. Res. Lett.* **36** L13201

[9] Farrell W M *et al* 1999 Detecting electrical activity from Martian dust storms *J. Geophys. Res.* **96** 11033–43

[10] Fabian A, Krauss C, Sickafoose A, Horanyi M and Robertson S 2001 Measurements of electrical discharges in Martian regolith simulant *IEEE Trans. Plasma Sci.* **29** 288–91

[11] Krauss C E, Horanyi M and Robertson S 2006 Modeling the formation of electrostatic discharges on Mars *J. Geophys. Res.* **111** E02001

[12] Farrell W M *et al* 2003 A simple electrodynamic model of a dust devil *Geophys. Res. Lett.* **30** 250

[13] Zhai Y *et al* 2006 Quasielectrostatic field analysis and simulation of Martian and terrestrial dust devils *J. Geophys. Res. Lett.* **35** 16

[14] Cummer S A 2000 Modeling electromagnetic propagation in the Earth ionosphere waveguide *Antennas Propag.* **48** 1420

[15] Eden H F and Vonnegut B 1973 Electrical breakdown caused by dust motion in low-pressure atmospheres: considerations for Mars *Science* **180** 962

[16] Hintze P E *et al* 2010 Alteration of five organic compounds by glow discharge plasma and UV light under simulated Mars conditions *Icarus* **208** 749–57

[17] Kok J F and Renno N O 2009 Electrification of wind-blown sand on Mars and its implications for atmospheric chemistry *Geophys. Res. Lett.* **36** 5

[18] Anderson M M *et al* 2012 The Allen Telescope Array search for electromagnetic discharges on Mars *Astrophys. J.* **744** 15

[19] Ondarkova A *et al* 2008 Peculiar transient events in the Schumann resonance band and their possible explanation *J. Atmos. Sol-Terr. Phys.* **70** 937–46

[20] Delory G T 2012 Problems and new directions for electrostatics research in the context of space and planetary science *Proc. 2012 Joint Electrostatics Conf.*

[21] Harrison R G *et al* 2010 Self-charging of the Eyjafjallojokull volcanic ash plume *Environ. Res. Lett.* **5** 024004

[22] Jackson T L *et al* 2010 Martian dust devil electron avalanche process and associated electrochemistry *J. Geophys. Res.* **115** E5

[23] Buhler C R, Mazumder M K, Calle C I, Clements J S, Srirama P K and Chen A 2008 Development of the dust particle analyzer *Proc. Pan Pacific Imaging Conf.*

[24] Lemmon *et al* 2004 Atmospheric imaging results from the Mars Exploration Rovers: Sprit and Opportunity *Science* **306** 1753–6

Chapter 9

The electrostatic environments of Venus and Mercury

9.1 Electrical phenomena in the Venusian atmosphere

Venus is in many ways very similar to the Earth. The two planets have about the same size, density, and chemical composition. However, the Venusian atmosphere is made up mostly of carbon dioxide (96.5% CO_2 and 3.5% nitrogen) with a density about 90 times that of the Earth and a temperature near the surface that reaches 730 K. This high temperature is due to the much denser atmosphere that traps 99% of the infrared radiation reradiated from the surface. This greenhouse effect is the most powerful in the Solar System.

The lower atmosphere is clear up to about 32 km. Above that, a thin haze of dust and aerosols reaches up to about 47 km, where the 20 km thick cloud layer that surrounds the planet starts. The lower part of this layer contains large solid particles that were discovered by the Pioneer Venus mission and that are yet of unknown origin. The outer part of the layer is made up of small sulfuric acid droplets. These droplets can acquire large electric charges and may generate lightning. Although no direct measurements of this charging mechanism have been performed, observations from terrestrial telescopes as well as with instrumentation on NASA's Pioneer Venus and ESA's Venus Express have provided strong evidence for lightning in the Venusian atmosphere. Russell and collaborators consider this evidence to be overwhelming [1]. However, some theoreticians are not convinced. Michael *et al*, for example, conclude that the expected electrical conductivity of the atmosphere and the expected abundance of aerosols do not support lightning on Venus [2].

Unlike Earth, Venus lacks an intrinsic magnetic field. Thus the solar wind particles interact directly with the upper atmosphere, since there is no magnetosphere to deflect them. UV radiation from the Sun ionizes atoms in the upper atmosphere, creating a highly conductive ionosphere. This ionosphere deflects the solar wind around the planet, forming a bow shock and a magnetotail [3]. The interaction of the

doi:10.1088/978-1-6817-4477-3ch9
9-1

Figure 9.1. Conceptual image of Venus Landsail, Zephyr, the NASA-proposed Venus lander. (Courtesy of NASA.)

solar wind with the top of the ionosphere produces a complex electrical environment that causes the ejection into space of charged particles in the magnetotail. Venus Express found that this environment also sends solar wind particles deep into the upper atmosphere, where they collide with the carbon dioxide molecules, producing the faint flashes of light that had been observed before. Data from the Venus Express magnetometer and the low-energy particle detector showed a plasma flow toward the planet that was 3400 km wide that lasted for 94 s [4].

Venus Express ended its mission in 2014. On 6 December 2015, Akatsuki, the Venus Climate Orbiter from the Japanese Space Agency, went into orbit around Venus. The craft is studying the atmosphere and its cloud layer, and is looking for additional evidence for lightning. NASA is now proposing to send a lander to Venus in 2024. The craft will have a relatively large sail and will travel on the surface powered by the slow winds of the dense Venusian atmosphere (figure 9.1). If approved, the Venus Landsail, Zephyr, will spend 50 days on the surface of Venus. Ground observations may provide direct evidence for lightning on Venus.

9.2 The electrostatic environment of Mercury

Mercury, the planet closest to the Sun, has an extremely thin atmosphere with a surface pressure of about 5×10^{-15} bar, which is essentially a vacuum. Mercury is the smallest planet in the Solar System, with a diameter of 4879 km, it is only slightly larger than the Moon (figure 9.2). Like the Moon, Mercury is exposed to the solar

Figure 9.2. Mercury from NASA's MESSANGER spacecraft, which studied the planet from 2008 to 2015. (Courtesy of NASA.)

wind, cosmic rays, and the Sun's UV radiation. The surface of the planet acquires an electric potential that balances all the current fluxes reaching the surface. There are multiple currents, ranging from the protons and electrons in the solar wind; alpha particles, protons, and electrons in cosmic rays; photoelectrons; to secondary electrons and GCRs partially reflected from the surface. Of those, only the solar wind protons and electrons, the photoelectrons, and the secondary electrons contribute most of the currents [5].

Mercury, like the Moon and all other airless bodies in the Solar System, is a charged body in a plasma. A photoelectron sheath forms on the side facing the Sun and a Debye sheath forms on the dark side. The solar wind flux is larger around Mercury, since it is much closer to the Sun, than it is around the Moon. Potentials on the surface should still be around +10 V on the day side of Mercury and in the negative hundreds of volts in the plasma wake of the night side, reaching higher values during solar storms.

A similar situation is expected around all moons without appreciable atmospheres. Titan and other moons with atmospheres should have different electrostatic environments dominated by their weather. Current and past exploration missions have yielded no evidence of atmospheric electricity on those bodies. But, as in the case of Venus, lightning is not easy to detect from orbit. Future missions may provide a better picture of the electrostatic environments of these bodies.

References

[1] Russell C T *et al* 2011 Venus lightning: comparison with terrestrial lightning *Planet. Space Sci.* **59** 956–73

[2] Michael M *et al* 2009 Highly charged cloud particles in the atmosphere of Venus *J. Geophys. Res.* **114** E4

[3] Zhang T L *et al* 2006 Magnetic field investigation of the Venus plasma environment: expected new results from Venus Express *Planet. Space Sci.* **54** 1336–43

[4] Zhang TL *et al* 2012 Magnetic reconnection in the near Venusian magnetotail *Science* **336** 567–70

[5] Stubbs T J, Halekas J S, Farrell W M and Vondrak R D 2005 Lunar surface charging: a global perspective using Lunar Prospector data *Workshop on Dust in Planetary Systems (Lunar and Planetary Institute)* 4070

Chapter 10

The electrostatic environments of the giant planets

10.1 The electrostatic and magnetic environments of Jupiter

Jupiter, the largest planet in the Solar System, has an atmosphere composed mainly of hydrogen (89.8% by volume) and helium (10.2%) gases, similar to that of the Sun. Its clouds of ammonia crystals, ammonium hydrosulfide crystals, and frozen water form the red, white, and brown bands visible from Earth even with small telescopes. The Great Red Spot is a giant storm larger than the Earth that has existed for hundreds of years.

At the lower layers of the atmosphere, the high pressure compresses the hydrogen gas into a liquid, forming an ocean around the entire planet. Deeper inside Jupiter, the huge atmospheric pressure dissociates hydrogen molecules in this vast ocean and produces an ionized plasma, making the liquid hydrogen electrically conductive [1]. The rapid rotation of the planet creates electrical currents that generate Jupiter's strong magnetic field (figure 10.1). This field engulfs the planet with an extensive magnetosphere that reaches out beyond the orbit of many of its 53 moons[1] on the side facing the Sun, extending into a long tail that reaches beyond the orbit of Saturn. Jupiter's rotationally driven magnetosphere derives its energy from the planet's rotation rather than from the solar wind, like the Earth's magnetosphere [2].

Jupiter's magnetic field traps large quantities of charged particles in belts, similar to Earth's Van Allen Belts. Due to the planet's rapid rotation, some of these charged particles are ejected into an immense current sheet 2–4 Jupiter radii in thickness that lies almost precisely in the magnetic equatorial plane [3]. The main source of plasma particles in Jupiter's magnetosphere is its moon Io, which contributes sulfur and oxygen ions. A second plasma source in Jupiter's magnetosphere is the solar wind,

---

[1] Jupiter has 53 confirmed moons and 14 recently discovered provisional moons which have a temporary designation by the International Astronomical Union.

doi:10.1088/978-1-6817-4477-3ch10

Figure 10.1. Artist's rendition of Jupiter's magnetic field lines. (Courtesy of NASA.)

which contributes protons and electrons. Although the mass density of the solar wind is much lower than the density of contributions from Io, the number density of protons in the solar wind reaching Jupiter's magnetosphere is comparable to that of Io [4]. Protons (H^+) and hydrogen molecule-ions (H_2^+) from Jupiter's ionosphere are the third plasma source in Jupiter's magnetosphere. Compared to the other two, the ionosphere is not a major source of plasma. Finally, sputtering of the surfaces of Europa, Ganymede, and Callisto, the three icy moons of Jupiter, contributes a small amount of plasma to the magnetosphere [5].

High-energy plasma particles can penetrate Jupiter's atmosphere near the magnetic poles and excite hydrogen and helium gases which in turn emit energy in the visible part of the spectrum, producing the largest and most energetic auroras in the Solar System. In 2016, astronomers started an observation program to observe and measure auroras on Jupiter with the Hubble Space Telescope Imaging Spectrograph (figure 10.2).

10.2 Lightning on Jupiter

NASA's Voyager 1, Galileo, and Cassini missions obtained optical images of lightning on Jupiter. Analysis of the images indicates that lightning occurs at a rate of 4×10^{-3} km^{-2} s^{-1}, much lower than the rate on Earth of 6 km^{-2} s^{-1} [6]. However, lightning on Jupiter appears to be much more energetic than on Earth. The average energy per bolt of lightning on Jupiter is about 10^{12} J, much larger than that on Earth, with an average energy per bolt of 10^9 J.

Besides visual and photographic detection of lightning bolts, lightning is detected on Earth with radio receivers that measure the radio frequency (RF) radiation emitted when lightning occurs. This RF peaks at a frequency of 10 Hz, in the very

Figure 10.2. NASA's Hubble Space Telescope image of an aurora near Jupiter's magnetic pole. (Courtesy of NASA.)

low frequency (VLF) region of the spectrum and extends to the extremely low frequency (ELF) region. Because these frequencies are reflected by the Earth's ionosphere and remain in the atmosphere, these radio waves are called *atmospherics*, which has been shortened to *sferics*. Terrestrial lightning also generates electromagnetic waves in the tens of hertz to tens of kilohertz (VLF to audible range). These waves can penetrate the ionosphere and travel along the Earth's magnetic field. Because they produce a gliding tone (tones that appear to glide up and down in pitch) when played through a loudspeaker, they are called *whistlers*. As these whistler waves interact with energetic electrons in the Earth's magnetosphere, they generate chorus and hiss waves (see chapter 5) [7].

Voyager 2 detected whistlers in Jupiter's atmosphere with the Plasma Wave Subsystem (PWS) instrument on the spacecraft [8]. Analysis of this data showed the lightning rate to be 4×10^{-3} km^{-2} s^{-1}, which agrees with the rate obtained from optical detection. However, no high-frequency sferics have ever been detected on Jupiter [9]. William Farrell at the NASA Goddard Space Flight Center thinks that since lightning discharges on Jupiter appear to be slower than on Earth by two powers of ten, the frequency of most of the emitted radiation would lie below the cutoff frequency required to penetrate the planet's ionosphere and were therefore not detected by any of the spacecraft [10]. Still, Farrell's model indicates that at five Jupiter radii, signal amplitudes from lightning discharges in the atmosphere should be of the order of a few $\mu V\ m^{-1}$, which are within the sensitivity of Voyager's PWS receiver. The discrepancy might be cleared by the new NASA Juno mission to Jupiter (figure 10.3).

Juno, which arrived at Jupiter in July of 2016, is scheduled to orbit the planet for 20 months and, at the end of the mission, deorbit and plunge into the atmosphere to prevent any contamination of Jupiter satellites. Among Juno's mission goals are to explore and study its magnetosphere near the poles to obtain new information on how the planet's strong magnetic field affects its atmosphere. Juno carries five sets of instruments to study Jupiter. The magnetometers will study Jupiter's deep structure by mapping the planet's magnetic fields; a micro-wave radiometer will probe Jupiter's deep atmosphere and measure the amount of water; the Jupiter Energetic-particle Detector Instrument (JEDI), the Jovian Auroral Distributions Experiment (JADE), and the Plasma Waves Instrument (WAVES) will sample electric fields, plasma waves, and charged particles around Jupiter to determine how the magnetic field links to the atmosphere; the UVS and JIRAM UV and infrared cameras will provide images of the electrical activity in the atmosphere and on the auroras; and the JunoCam will take close-up photo-graphs of Jupiter.

Figure 10.3. Artist's concept of the Juno spacecraft as it arrives at Jupiter. (Courtesy of NASA.)

10.3 The electrostatic environment of Saturn

Saturn is in many aspects very similar to Jupiter. It has an atmosphere composed of about 75% hydrogen and 25% helium with traces of other gases, with clouds of ammonia crystals, ammonium hydrosulfate crystals, and frozen water. Deep in the atmosphere, the hydrogen gas turns into an electrically conductive liquid. As the planet rotates, electric currents are formed in this metallic liquid that in turn generate a magnetic field. The interaction of the solar wind particles with Saturn's magnetic field forms its magnetosphere. NASA'S Cassini mission to Saturn, which carried a Magnetosphere Imaging Instrument (MIMI), produced an image of Saturn's magnetosphere in 2004 by detecting the hydrogen atoms emitted from regions well outside Saturn's rings (figure 10.4).

The Voyager 1 mission detected chorus and hiss waves as well as RF signals with its Planetary Radio Astronomy (PRA) instrument, but its imager detected no optical flashes [9]. The PRA on both Voyager 1 and 2 also detected unusual radio discharges of short duration and narrow band with frequencies ranging from 20 kHz to 40 MHz, the upper frequency limit of the PRA [11]. These signals were named Saturn electrostatic discharges (SEDs). These signals originate in the equatorial region of Saturn's atmosphere. In 2004, the Radio and Plasma Wave Science (RPWS) instrument on Cassini detected about 5400 SEDs, which were likely correlated with a possible large equatorial lightning storm lasting several months [11]. The difficulty with establishing this correlation was due to the lack of whistler detection. Neither the two Voyagers nor Cassini had been able to detect these dispersive signals, in spite of the fact that they are the source of the observed chorus

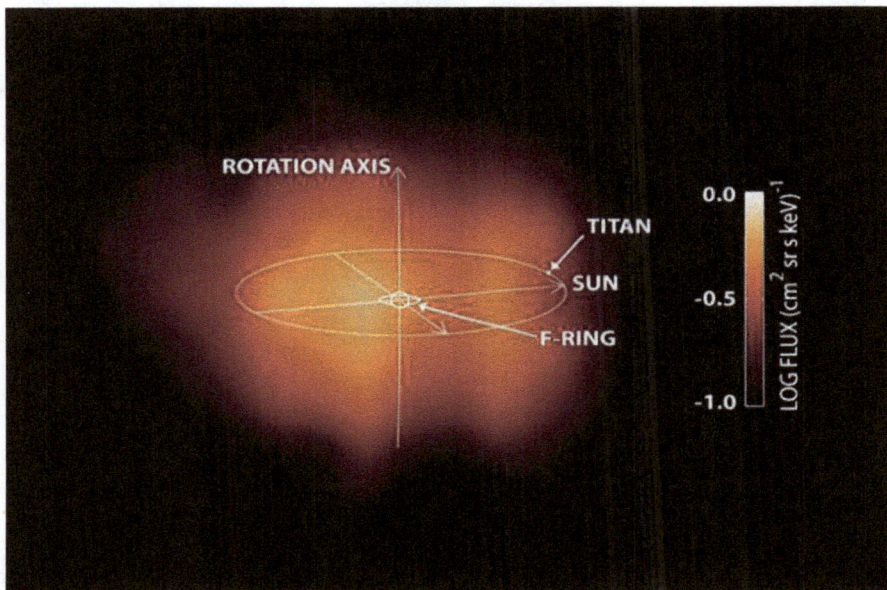

Figure 10.4. Image of Saturn's magnetosphere taken by Cassini's MIMI on June 21, 2004 from a distance of about 6 million km. (Courtesy of NASA.)

and hiss waves. Whistler waves propagate along magnetic field lines. Voyager 1 and 2 approached within 3.1 and 2.7 Saturn radii, respectively, thus crossing the magnetic field lines. If the storm was localized in the equatorial region, as suspected, the Voyagers would not have crossed the field lines along which the whistlers traveled [9], and Cassini was far outside the magnetosphere and could not detect these signals [11]. A direct observation of lightning was needed.

The first direct visual observation of lightning on Saturn was obtained in August 2009 with Cassini's Imaging Science Subsystem from a distance of 35.5 Saturn radii [12]. The images were taken on Saturn's night side, six days after the planet's equinox, when the Sun shone directly over the equator, with the rings on edge to the sunlight which minimized light from the rings shining on Saturn. The optical energy of a single flash was calculated to be comparable to that of lightning on Earth and Jupiter, with values up to 1.7×10^9 J [12].

Cassini detected daytime lightning on Saturn while observing a giant storm on 6 March 2011 (figure 10.5). The storm began in December 2010 and lasted until June 2011. The size of the lightning flashes was measured to be about 200 km in diameter, which was similar to those detected in 2009. From this size, the Cassini mission investigators estimated that lightning occurs 125–250 km below the cloud tops, probably in the water clouds [13]. The optical energy of single flashes for the 2010–2011 storm was calculated to be 8×10^9 J, somewhat larger than the energy of the

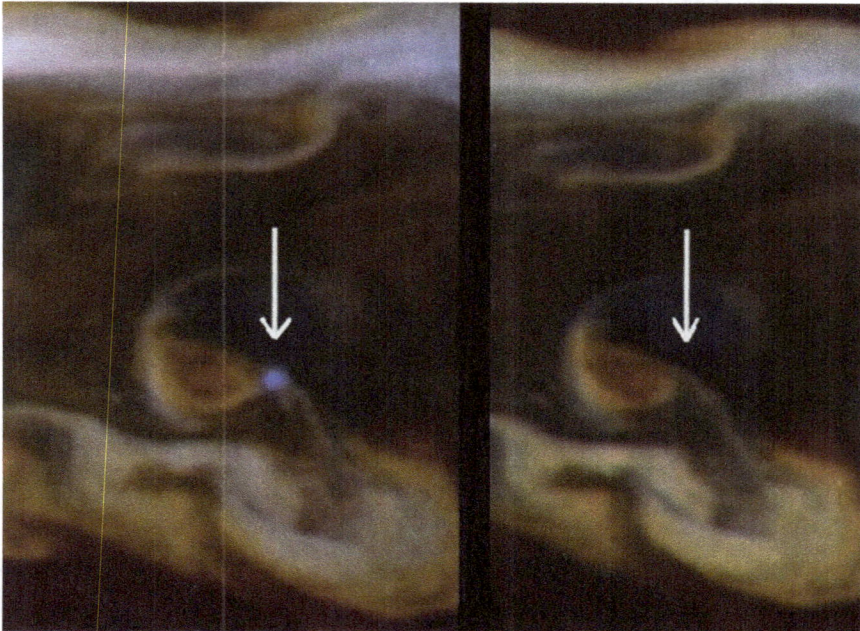

Figure 10.5. False color mosaics from NASA's Cassini spacecraft show lightning strikes during the giant 2010–2011 storm that engulfed the planet's northern hemisphere. The blue dot on the left mosaic is the lightning strike. The mosaic on the right was taken 30 min later and the strike has disappeared. (Courtesy of NASA.)

2009 storms. Radio energies were estimated to be of the same order of magnitude as optical energies.

The total power of the 2010–2011 storm was calculated at about 10^{17} J, which is in the same order of magnitude as the total power that Saturn radiates to space. Dyudina and collaborators conclude that these giant lightning storms play an important role in Saturn's cooling and thermal evolution [13].

10.4 The electrostatic environments of Uranus and Neptune

Voyager 2 has been the only spacecraft to fly close to Uranus, passing within 81 500 km of the planet in January of 1986. Voyager 2 discovered that Uranus has a strong magnetic field that is inclined at a 59 degree angle to the planet's axis of rotation and has its center displaced by about 30% of the planet's radius. The inclination of Uranus's magnetic field is larger than those of Saturn, Jupiter, Earth, and Mercury, which are all within about 12 degrees of their axes. Due to this large tilt, the magnetic field lines are wound in a corkscrew shape. The source of Uranus's magnetic field is unknown. No electrically conductive liquid deep in the atmosphere that could produce a magnetic field has been discovered.

Uranus's magnetosphere has a tail that extends to 10 million km. The plasma density of the magnetosphere is very low, about 0.1–1 charge particle (protons or electrons) per cubic centimeter [14]. Voyager discovered two radiation belts around Uranus containing mostly protons.

Voyager 2's PWS instrument detected electrostatic discharges from Uranus that are similar to those detected from Saturn. These Uranus Electrostatic Discharge (UED) events are not as intense as SED events and are also less frequent. It is not known if these UEDs are caused by lightning.

Voyager 2 is also the only spacecraft to visit Neptune, flying within 5000 km of the planet in the summer of 1989. The spacecraft measured Neptune's magnetic field and mapped its magnetosphere. The planet's magnetic field is tilted 47 degrees from Neptune's axis of rotation and is displaced 55% from the planet's geometrical center. Voyager detected auroras over wide regions of the atmosphere of Neptune which had an intensity of about 50% of those on Earth. It also detected whistler waves.

Neptune has a high-altitude cloud layer of methane, a hydrogen sulfide–ammonia (H_2S–NH_3) cloud layer, and deeper layers of water and ammonium hydrosulfide. Lightning could take place in some of these dense cloud layers [15]. Despite the existence of whistlers, no optical confirmation of lightning activity has been obtained. Gibbard and collaborators used a particle-growth and charge separation model that had been applied with great success to both the Earth and Jupiter to investigate the possibility of lightning in the cloud layers of Neptune [15]. With this model, they found that if charge transfer during non-sticking particle collisions can take place faster than 1% of the rate between water ice particles, lightning may happen at the maximum mass densities predicted for the H_2S–NH_3 cloud layer. The model predicted that lightning is inhibited in the deep water and ammonium hydrosulfide cloud layers.

Uranus and Neptune are the only two planets in the Solar System that have not been studied from an orbiting spacecraft. NASA is looking into a possible mission to these two planets in the late 2020s or early 2030s. When these missions take place, new data on the cloud structure and other atmospheric properties of these two planets will allow scientists to determine the nature of electrical phenomena in these atmospheres.

10.5 The electrostatic environment of Saturn's moon Titan

Titan is Saturn's largest satellite. We include it here with the giant planets because it is part of the Saturnian system and because it is the only satellite in the Solar System with a substantial atmosphere. Titan's atmosphere is composed of 96.8% nitrogen, 3% methane, and 0.2% hydrogen with traces of hydrocyanic acid, ethane, propane, acetylene, ethylene, and other gases. The atmosphere has a surface pressure of 1.6 bars. Knowledge of the composition of Titan's atmosphere after Voyager 1's 1980 flyby encouraged scientists to suggest that lightning may exist on the satellite. Voyager's instruments did not detect any radio emissions indicative of lightning on Titan [9], but laboratory experiments show that many of the gases in Titan's atmosphere could be produced by electrical discharges [16]. Although these compounds can also be made with other chemical processes, the abundancy of ethylene is higher than what is expected to be produced by these other processes.

In 2001, Tokano and collaborators proposed a model for the generation of lightning on Titan due to cloud electrification during collisions caused by rapid vertical motion of solid methane particles in the atmosphere [17]. According to this model, free electrons produced in cosmic ray ionization or impacts from energetic magnetosphere particles attach to methane particles in the cloud. When these negatively charged particles approach Titan's water ice surface, they may form an image charge. This charge separation may generate a maximum temporary electric field of 2.5×10^6 V m^{-1}, large enough to overcome electrical breakdown and produce lightning in the lower atmosphere.

The Cassini mission to the Saturnian system has conducted over 100 flyby passes of Titan since 2004. The 118th flyby occurred on 25 March 2016, passing within 980 km of Titan's surface to observe the atmosphere with the Ultraviolet Imaging Spectrograph (UVIS) and the Ion Neutral Mass Spectrometer (INMS). Data from the UVIS will help determine the composition, distribution, and particle content of the atmosphere. INMS data are used to determine the composition of positive ions and neutral particles in the upper atmosphere. These observations may be used to check and improve models such as Tokano's.

Cassini's RPWS instrument has not detected radio emissions associated with possible lightning in Titan's atmosphere [18]. The European Space Agency's Huygens probe, which was released from NASA's Cassini spacecraft and descended to the surface of Titan in January of 2005, carried the Huygens Atmospheric Structure Instrument (HASI) and the Permittivity, Wave, and Altimetry (PWA) instrument to detect radio signals from lightning (figure 10.6). HASI-PWA detected an unusual ELF signal at 36 Hz [19]. After several years of analysis of the data

Figure 10.6. Artist's rendering of the descent and landing of the Huygens Probe. (Courtesy of NASA/ESA.)

collected by these two instruments, Beghin and collaborators argued that this ELF signal displays the characteristics of a second harmonic of a Schumann resonance [20]. On Earth, Schumann resonances are generated by lightning in the region (known as *cavity*) between the Earth's surface and the ionosphere. However, the acoustic sensor on Huygens did not detect thunder that could be associated with lightning [19]. Beghin and collaborators have proposed an alternative powering mechanism. They suggest that the interaction of Titan's ionosphere with Saturn's magnetosphere generates electrical currents in the atmosphere, producing a dynamo effect when the magnetosphere plasma rotates with Saturn. The lower boundary of Titan's cavity is proposed to be a conductive water–ammonia ocean residing 55–80 km below an ice crust. These ionospheric current sources and not lightning produce the observed Schumann resonance.

References

[1] Pierleoni C *et al* 2016 Liquid–liquid phase transition in hydrogen by coupled electron–ion Monte Carlo simulations *Proc. Natl Acad. Sci. North Am.* **113** 4953–7

[2] Khurana K K *et al* 2004 The configuration of Jupiter's magnetosphere *Jupiter: The Planet, Satellites and Magnetosphere* ed F Bagenal, T E Dowling and W B McKinnon (Cambridge: Cambridge University Press)

[3] Goertz C K 1976 The current sheet in Jupiter's magnetosphere *J. Geophys. Res.* **81** 3368–72

[4] Hill T W *et al* 1983 Magnetospheric models *Physics of the Jovian Atmosphere* ed A J Dressler (Cambridge: Cambridge University Press)

[5] Cooper J F *et al* 2001 Energetic ion and electron irradiation of the icy Galilean satellites *Icarus* **149** 133–59

[6] Uman M A 1987 *The Lightning Discharge* (Orlando, FL: Academic Press)

[7] Shawhan S D 1979 *Solar System Plasma Physics* ed C F Kennel *et al* (Amsterdam: North-Holland) p 221

[8] Gurnett D A *et al* 1979 Plasma wave oscillations near Jupiter: initial results from Voyager 2 *Science* **206** 987

[9] Desch S J *et al* 2002 Progress in planetary lightning *Rep. Prog. Phys.* **65** 955–97

[10] Farrell W M 2000 *Radioastronomy at Long Wavelenghts* ed R G Stone *et al* (Washington, DC: American Astrophysical Union)

[11] Fischer G *et al* 2006 Saturn lightning recorded by Cassini/RPWS in 2004 *Icarus* **183** 135–52

[12] Dyudina U A *et al* 2010 Detection of visible lightning on Saturn *Geophys. Res. Lett.* **37** L09205

[13] Dyudina U A *et al* 2013 Saturn's visible lightning, its radio emissions, and the structure of the 2009–2011 lightning storms *Icarus* **226** 1020–37

[14] de Pater I and Lissauer J J 2010 *Planetary Sciences* 2nd edn (New York: Cambridge University Press)

[15] Gibbard S G *et al* 1999 Lightning on Neptune *Icarus* **139** 227–34

[16] Fujii T and Arai N 1999 Analysis of N-containing hydrocarbon species produced by a CH_4/N_2 microwave discharge: simulation of Titan's atmosphere *Astrophys. J.* **519** 858–63

[17] Tokano T *et al* 2001 Modelling of thunderclouds and lightning generation on Titan *Planet. Space Sci.* **49** 539

[18] Fischer G and Gurnett D A 2011 The search for Titan lightning radio emissions *Geophys. Res. Lett.* **38** L08206

[19] Grard R *et al* 2006 Electric properties and related physical characteristics of the atmosphere and surface of Titan *Planet. Space Sci.* **54** 1124–56

[20] Beghin C *et al* 2012 Analytic theory of Titan's Schumann resonance: constraints on ionospheric conductivity and buried water ocean *Icarus* **218** 1028–42

www.ingramcontent.com/pod-product-compliance
Lightning Source LLC
Chambersburg PA
CBHW082111210326
41599CB00033B/6661